THE WASTE AND THE BACKYARD

Environment & Management

VOLUME 8

The titles published in this series are listed at the end of this volume.

The Waste and the Backyard

The Creation of Waste Facilities: Success Stories in Six European Countries

Edited by

BRUNO DENTE and PAOLO FARERI
Istituto per la Ricerca Sociale, Milano, Italy

and

JOSEE LIGTERINGEN
Center for Clean Technology and Environmental Policy, University of Twente, Enschede, The Netherlands

Springer-Science+Business Media, B.V.

A C.I.P. Catalogue record for this book is available from the Library of Congress.

Printed on acid-free paper

ISBN 978-90-481-5021-2 ISBN 978-94-015-9107-2 (eBook)
DOI 10.1007/978-94-015-9107-2

Originally published by Kluwer Academic Publishers in 1998.
Softcover reprint of the hardcover 1st edition 1998

Contents

PREFACE

This book presents the results of a research project on the siting of waste facilities that was carried out during the period 1992 until 1994 in seven European countries. In France, Hungary, Italy, The Netherlands, Slovenia, Spain and Switzerland a specific case of the siting of a waste incineration plant was selected according to previously set criteria in order to study the decision making process. The aim of the project was to investigate the decision making processes on the siting of these particular waste facilities in order to identify the factors that can explain and predict the success of the decision making process. A theoretical framework was used to analyse and compare the decision making processes in the selected cases. This framework, based on a public policy analysis approach, was developed at the Instituto per la Ricerca Sociale by Bruno Dente and Paolo Fareri.

The research project was funded by the European Union Directorate General XII, in the RTD Program in the Field of Environment, research area III: Research on Economic and Social Aspects of Environmental Issues. The Swiss study was financed by the Federal Office for Education and Sciences out of the special fund reserved for research projects conducted within the frame of European research programmes of the European Union.

The comments of Angela Liberatore were very helpful in carrying out the research work. In the beginning the research project was carried out only in five European countries. Later on an extension of the project to two Eastern European countries was made possible by the European Union. This allowed us to have a Slovenian and a Hungarian case study in the project and gave us the opportunity to study two decision making processes on siting a waste facility in Eastern European countries.

The seven selected cases were studied by the Universitad Autonoma de Barcelona (Spain), the Laboratoir d'Observation de l'Economie et des Institutions Locales (France), the Center for Clean Technology and Environmental Policy of the University of Twente (The Netherlands), Institut de Hautes Études en Administration Publique (Switzerland), the University of Ljubljana (Slovenia) and the IEM Institute for Environmental Protection (Hungary). Because of the fact that the Hungarian siting procedure was not completed at the time this book went to press, an analysis of the Hungarian case is omitted.

The research work was made possible largely thanks to the co-ordination by the Instituto per la Ricerca Sociale. The editing of the final research report for publication is partly due to the contributions of the Center for Clean Technology and Environmental Policy of the University of Twente (The Netherlands) and the Instituto per la Ricerca Sociale (Italy). In particular we would like to thank Hans Bressers for his editorial suggestions. Finally, we would like to thank the authors of this book for their willingness to revise their chapters more than once in order to achieve coherence.

It is our hope that the book will be interesting to readers in the environmental field as well as in the field of policy studies. It is not meant to compare decision making processes on waste facilities siting in different countries, but to derive lessons from success stories of the decision making process of waste facilities that happened to be sited in six different European countries.

Bruno Dente
Paolo Fareri
Josee Ligteringen

1 SITING WASTE FACILITIES: DRAWING LESSONS FROM SUCCESS STORIES

BRUNO DENTE, PAOLO FARERI

1.1 LEARNING FROM SUCCESS

Controversies about the siting of facilities for the disposal and treatment of both hazardous and domestic waste are widespread in all industrialised countries. Proponents (both public and private) of these facilities point to the need for them and to their technical suitability, while opponents (local communities and environmental groups) emphasise the risks for health and environment associated with the substances to be disposed of or treated by the process.

Starting in the early '80s, most of these controversies resulted in the abandonment of the proposed projects; as a consequence the volume of waste 'waiting' for disposal and treatment is growing, while old, leaky landfills are still being used. The paradoxical situation of projects which are needed for environmental reasons and which are opposed on environmental grounds has been addressed by scholars and by policy makers as well, and solutions have been looked for.

The causes of the NIMBY (Not In My Back Yard) syndrome have been enquired into in policy analysis, risk analysis and the sociological literature. The most accurate studies argue that it is not only the characteristics of the projects (their technical suitability, economic efficiency, etc.) but also, or even more so, the features of the decision making process that are crucial in determining the acceptability and feasibility of waste facilities projects.

Possible remedies for the NIMBY syndrome have been suggested and some of them, like the introduction of some information and participatory procedures in the decision making process (the Environmental Impact Assessment procedure being a case in point), have

B. Dente et al. (eds.), The Waste and the Backyard, 3–46.

been implemented. However, waste disposal facilities were actually built and are operating in only a few cases.

Within this context, the aim of the research project whose results are presented here was to investigate decision making processes for the creation and siting of waste facilities in order to identify the factors that are able to predict the success of the projects under consideration.

Adopting a Public Policy Analysis approach, seven successful cases have been selected in seven European Countries: France, Italy, the Netherlands, Spain, Switzerland, Hungary and Slovenia. An in-depth analysis of the decision making processes leading to a positive outcome have been conducted, reconstructing the course of events from the posing of the problem to the building of the facility, identifying the participating actors and their behaviour, and analysing the interactions between them. The objective of this analysis is to identify, in each case study, those factors related to the characteristics of the decision making process that are responsible for success. On this basis, and through the comparison of the case studies, we try to identify success factors that are verified in different contexts.

The general hypothesis adopted here is that the improvement of the effectiveness of public policy making depends on the capacity to learn from the decision making processes as they actually occur, and to establish the conditions for the diffusion of the factors that have been identified - in a specific situation - as responsible for success. Following this hypothesis we try to draw some lessons from case study analysis, by giving some indications for the redefinition of public policy making in the field of waste treatment, in order to maximise the chances of reaching a positive outcome.

The interest in the issue of waste facilities siting is twofold, from the Policy Analysis point of view. One could say that understanding success is important because of the relevance of the specific issue being treated, that is, for what concerns the content of the decision making processes considered. On the other hand, the relevance of the results of the analysis goes beyond the issue of waste facility siting. As we will see later, the more general question of how success can be achieved 'under adversity' is treated here. Some of the conclusions could then be applied to other fields of public policy making characterised by the same problems.

The decision to concentrate attention on success stories needs an explanation, since most of the existing literature is dedicated to the analysis of failure. The study of failure developed within an approach based on the assumption that a general (that is, applicable in the same way to different contexts) 'model' for success can be identified. Within this frame the factors (the independent variables) for success are known: they are included in the ideal model and they are a condition for the analysis. The objective of studying failures is then to identify the success factors whose absence - in the specific situation under study - is responsible for the 'negative' outcome of the process. As a consequence, the lessons that can be drawn - for what concerns the improvement of the effectiveness of public policy making - from such an approach are about the ways one could identify to facilitate the diffusion of the 'ideal' success factors that are lacking.

The methodological approach we are proposing here is based on different assumptions. The general hypothesis is that it is impossible to define *ex ante* a single, general model for success. An ideal 'one best way' does not exist. And, consequently, we do not have a model to be confronted with the decision making processes taken into consideration in order to understand failure, because we do not know - before the analysis - what the success factors are. A corollary of this hypothesis is that the supposed identity of factors for success and factors for failure (the latter being actually the absence of the first) is not valid. Factors for success and factors for failure are different.

The accumulated experience demonstrates that many factors for failure can usually be identified, which are often closely linked with the contingent aspects of the processes analysed. This reduces the possibility of deducing general laws from the empirical analysis. But, apart from that, studying cases of failure eludes the real methodological question, which is the need for an analytical reconstruction of a 'model' for success that does not exist in 'ideal' terms. In order to do that we have to change the field of analysis, from that of failure to that of success. This is why success stories, within this approach, represent the focus of the analysis. The aim of the research is then to identify the factors responsible - in the specific decision making processes taken into consideration - for the positive outcome, in order to understand how these factors could be reproduced and/or diffused in a different situation, establishing the conditions that facilitate success.

In short, while the 'traditional' approach tends to explain why - in general - the considered processes fail, the one proposed here tries to explain why - in particular - a positive outcome is reached, and on this basis to draw general lessons for improving the effectiveness of public policy making.

As success represents the focus of our analysis, a clarification is needed about its definition. Success is actually an ambiguous term, used in different contexts with different meanings. *Success* is defined here as *the possibility to build and operate a facility in a reasonable time within a given geographical framework.* This in turn implies the possibility of finding an agreement between all the parts involved. In other words, the suggested definition of success does not try to determine what is right and what is wrong in absolute terms; rather it tries to establish whether a solution to the siting problem is found which is suitable for a given environmental and social context.

This definition of success is grounded on two main reasons. The first one has to do with the main *objective* of the research. We are focusing on the performance of policy networks, on the identification of the factors facilitating the solution of the typical constraints of these decision making processes, and not on the definition of the 'best' solution to waste treatment. The second one, directly related to the first, has to do with the *field of analysis* of the research. The case studies have been selected by looking at one specific solution - the realisation of a waste facility - that is not necessarily, in general, the best solution for waste treatment. It is, by the way, one of the most difficult to implement. We will come back to this question in the presentation of the results (see section 1.3). Success is then related to the decision making process, and not to the problem of waste in general.

Just a few words about case study selection in the countries considered: the same criteria have been adopted for selecting the case studies. In general we have been able to find a decision making process that corresponds to almost all the criteria identified: process termination (e.g. the existence at least of an agreed siting decision, if not the presence of an already operating facility); recently proposed projects; processes preferably regarding the siting of hazardous waste treatment plants; process complexity (where the presence of conflict has been considered as an indicator of complexity, and not as a condition for selection, another indicator being the presence of a public-private partnership for the realisation of the facility); use of complex technologies for waste treatment.

It should be noted, on the other hand, that in most countries the selection has been made among a very limited number of potential case studies. The research work done by the national teams for case study selection demonstrates that success in waste facilities siting is actually very difficult to achieve. In particular, it has been very difficult to find cases of success related to recently proposed projects: most of the existing facilities in the countries considered were implemented quite a long time ago (especially during the '70s).

1.2 PUBLIC POLICY ANALYSIS AND WASTE FACILITY SITING

A common theoretical framework for case study analysis (presented in Chapter 8) has been used by the national teams. A particular emphasis has been put on the need to adopt the same methodology, in order to maximise the opportunity offered by the project, to compare seven case studies in seven different European countries. The framework focuses on the identification of the relevant actors, their resources and their patterns of interaction, in order to explain why a certain decision has been taken. Starting from the definition of the analytical elements of public policies, it describes the method and techniques to be adopted for case study analysis, identifying the different phases, the related objectives, the instruments and the sources.

While the framework is designed to be applicable to different fields of study (and this is why its description is posed at the end of this volume), one of the efforts the group made - before starting with the analysis of the selected cases - has been to formulate hypotheses about the main characteristics of decision making processes concerning waste facilities siting. We tried to identify the specific aspects of these processes, as we expected them to be on the basis of the existing literature, and to formulate hypotheses about the possible factors responsible for the final positive outcome.

Referring to the general analytical elements described in the framework, these hypotheses concern: the situative variables, the policy problem, the stake, the structure of policy network and the patterns of interaction and, finally, actors' resources.

Before moving to the description of the general results (in section 1.3.), it may be useful to summarise these hypotheses, which represent the point of departure in the reflection the research group has made concerning the identification of success factors in waste facilities siting.

Situative variables. Within the factors that play a role in decision making, and in particular in decision making with respect to waste facilities siting, we can distinguish factors that are part of the context of the decision-making process: the so-called situative variables. The intervening variables that are linked to the situation are those elements that, not being part of the decision-making process, because the actors cannot modify them, can nevertheless modify actors' objectives and resources. Here we list some of these situative variables, realising that it is not a complete list.

Possible intervening situative variables could include *the social and economic conditions* of the country and/or the area. In this respect we could think for example of the unemployment rate. The building and functioning of a waste facility can bring considerable employment to a region. In regions that already have high unemployment rates, especially the contribution of employment resulting from siting a waste facility can be substantial, whereas the creation of employment in regions where employment is high already is less a deciding factor in agreeing on the siting of a waste facility. In this respect we may expect that it will be easier to compensate the local costs deriving from the realisation of a waste facility in an area characterised by negative occupational trends.

If we think of context variables that may influence the decision-making process, we could also think of the *degree of environmental consciousness*. The degree of environmental consciousness at the local level is particularly important. We may expect that in cases where the degree of environmental consciousness is very low the people of that local community value the siting of a waste facility in another way than in cases where the degree of environmental consciousness is high. The perception of waste disposal as a collective problem is an especially important component of environmental consciousness.

Another context variable can be found in the *degree of technological and scientific culture* at the societal level which can modify the perceptions about the risks involved in the proposed facility. Here we have to note the role of the media in influencing the perceptions of the risks of certain technological processes. Whereas the perception is influenced by the amount of information people receive about these processes, the information can be presented in such a way that it influences the perception in certain directions. We may expect that a community that perceives the risks involved as rather low reacts in another way to the siting of a waste facility in its backyard than a community that is characterised by the perception of very serious risks.

Furthermore we can think of context variables like the *occurrence of specific events*, like accidents with grave environmental impacts. The occurrence of such an event can substantially increase the emotional approach to the problem by some of the actors involved. This may lead to an intensification of certain conflict relations, or to the fact that 'silent' actors take the initiative of participating in the decision making process in order to present their point.

Finally we distinguish two more general situative variables, that can play an important role in the siting of waste facilities. One more general situative variable we distinguish is the *changing structure of institutional arrangements*, interfering with the distribution of resources and/or with the structure of the network. Where the distribution of resources or the structure of the network changes, the actors' power bases change. This can create new positions for new actors with new objectives. Here we can also think of coalitions that are created within the scope of other processes, and that are therefore also applied in the present decision making process.

Another more general situative variable we distinguish is the *definition, implementation or outcome of other policies* that, as a situative variable, may influence the decision making process. This variable can interfere with the decision making process by modifying the actors' objectives and/or resources, or the policy network: even factors that are not part of the decision making process can explain in some way the developments in decision making.

Specific aspects of the problem. Besides situative factors we can distinguish specific aspects of the problem at hand that influence the decision making process, in particular with respect to the siting of waste facilities. The realisation of a waste treatment facility can be seen by the relevant actors as a way of solving different types of problems, and in that way contributing to different types of objectives. In a first way the specific decision we are considering can be part of the implementation of a more comprehensive plan, including other facilities to be realised somewhere else (contributing to a general objective). Or, it can be the solution to the waste disposal problem of a specific area, without the possibility of externalising the costs (fulfilling a specific objective). Or again, the problem can be to increase the capacity of an already existing waste disposal facility.

Our hypothesis is that successful decision making will be easier if the problem definition is the outcome of an interaction including most of the actors involved, and in particular the potential opponents. Participating in the problem formulation involves the recognition of the existence of a 'problem' and, consequently, of the necessity to solve it. This should facilitate an approach by the opponents based on the reduction of costs, but in any case oriented to finding an agreement. Furthermore, in this way we can be sure that the actors at least have the same problem definition. It is easier to start discussing possible solutions from a shared problem definition than from unknown problem definitions.

Another relevant aspect has to do with the identification of the field of analysis (see Chapter 8): by defining the policy problem, the analyst actually identifies the boundaries of the decision making process, that is, our unit of analysis.

In defining these boundaries it is important to take account of the possible interdependence between success and failure. We may expect that in some cases the failure of a certain proposal for a waste facility is a condition - for instance in terms of redefinition of the behaviour of some of the actors involved - for the success of another one in a nearby location (the one we are analysing). In a case like this it is probably better to define the boundaries of the policy in order to include both the failure and the success as parts of the same decision making process and consequently extending the analysis to the first, unsuccessful attempt to build the facility. From a slightly different point of view we can say that the geographical area we are interested in, besides the specific location of the facility, is the one where the waste to be treated is produced, and not only the specific location of the facility that is finally decided upon after several attempts.

Issues about the stake. Besides specific situative variables and problem issues we distinguish issues about the stake that are typical of decision making on waste facilities siting, which can contribute to the explanation of the outcome of the decision-making process. We identify two elements related to the stake that should probably contribute to determining the policy outcome in the case of waste facilities siting. The first one is related to the *availability of alternative solutions*. The possibility to choose among alternatives can contribute positively to the final success only if it is linked to the possibility of fragmenting the solution. For example, if the project can be divided into differently located parts, this would contribute to a distribution of costs among the actors, facilitating an agreement. The availability of alternative locations is, in this case, a success factor. By contrast, if it is impossible to distribute costs among the actors, it can be easier to decide if there are no alternatives at all. This will put the actors under pressure in a positive way, making it clear that the only chance to solve the problem is to find an agreement on the only available solution, letting costs fall on just one actor. However, we have to note here that compensation is still possible, so that other actors may also experience costs in compensating the actor that bears the predominant costs of the decision making process.

This is how we get to the second element that is related to the stake: the *presence of compensation opportunities*. Decision making processes about waste facilities siting are characterised by the concentration of costs (on the residents of the nearby areas) and the broad distribution of benefits. In this situation an important success factor is the *possibility* to compensate the actor that is bearing the costs. This can be done by increasing the stake.

We have to notice two issues in regard to compensation opportunities. The first concerns the *distance* between waste producers and the waste facility. Let us take as an example the case of a big urban settlement producing waste. If the facility is located within the settlement, the situation is one where the same people who are profiting from treating the waste they produce are also suffering due to the operation of the facility. In this case the need for compensation should be less, because we have a balanced situation where costs and benefits fall predominantly to the same people. If, on the other hand, the facility is located far from the area where the waste is produced, we have a very unbalanced distribution of costs and benefits. In this case the need for compensation seems crucial for the success of the decision making process. However, this does not mean that a compensation arrangement is needed. Compensation can be included in the chosen option: for example by bringing employment to the region.

The second issue concerns the *origin* of compensation. Success of the decision making process should be easier if compensation comes from the actor that is benefiting from the facility. In our previous example this means compensation by the actor that produces the waste, but is not suffering because the facility is located far from him. In this case both actors pay (the actor that sees itself confronted with a waste facility in its backyard, as well as the actor that produces the waste), albeit in different forms. By contrast, success should be more difficult if the compensation is paid by a third party, for example by the State. In this case compensation only partially reduces the unbalanced distribution of costs and benefits: some actor remains - the one that produces the waste - to profit from the facility without paying costs.

In short, the hypothesis here is that, in the presence of compensation opportunities, who *pays* is more important than who is *refunded*.

Issues about the patterns of interaction. Besides the variables mentioned above, which can explain the developments within decision making on the siting of a waste facility, we can distinguish specific variables that are related to the interactions within the network of relevant actors.

Within the field of deciding on the siting of waste facilities we will probably find policy networks characterised by crowdedness, by heterogeneity of actors (different bodies of the public administration, private actors, social actors, political parties, etc.), and by the presence of 'ad hoc' actors. But, apart from these characteristics, three hypotheses can be proposed about the success factors related to another character: the pattern of interaction.

The first hypothesis has to do with the *level of polarisation of the network*. At a first glance the chances of success should be serious as the process is less polarised. It seems easy to get actors on one line if they already are near that line. This means that actors can express their objectives at an early stage, and consequently can discuss solutions to the problem at a moment in decision making when they still have many options. Where this is the case, we have to notice that a network may *seem* rather unpolarised, but in fact be very polarised. We can illustrate this, taking the different phases of the process into consideration. In the presence of potential conflicting interests, these interests may remain hidden until the process is in an advanced state of development. Experiencing these conflicting interests in a late stage of the decision making process can make it much more difficult to find an agreement than experiencing conflicting interests right from the start. If conflicting interests are present, the chances of a successful decision making are better if we have a polarised network right from the start, as this is a condition for the expression of the objectives by the actors and, consequently, for facing the problem when it is easier to modify the solution. So we could expect a shift from confrontation to negotiation as a possible road toward success. In short, a high level of polarisation can easier turn into a low level of polarisation, facilitating successful decision making, if it is recognised during an early stage of the decision making process.

A second hypothesis concerning the interactions within the network is related to the actors' roles. Because in cases of siting a waste facility we frequently have to do with a high rate of conflict, the *presence of a mediator* is a very important element contributing to the success of this type of process. The solution to a conflict situation often depends on the fact that the counterparts are not interacting directly, but through an actor, holding prevailing process objectives, getting benefits only by playing this specific role in the process or if in general a solution to the conflict is found, whoever wins or loses.

Finally, we can identify success factors related to the interactions within the network that concern the *level of formalisation of the rules of the game*. In particular in complex cases like the decision making on waste facilities siting, we could hypothesise that the more the rules of the game are strictly determined before the beginning of the process, for example in terms of legal procedures, the more it will be difficult to reach a positive outcome, because in this situation actors may not be allowed to interact in a flexible way, discussing possible options, and distributing costs and benefits according their own agreements. Determining the rules of the game as a part of the decision making process can precisely be identified as an important success factor. In this way determining the rules of the game can be an outcome of the interaction among the participants. In short, taking a successful decision often requires a flexible approach to the rules of the game from the start of the process.

Issues about actors' resources. Proceeding with our attempt to highlight the specific aspects of the decision-making processes about waste facilities siting, a relation can be established between the different roles of actors and the resources they need to succeed in the interaction with each other. Some relationships that seem to be present in decision making on the siting of waste facilities are presented below.

Concerning the actors playing the role of *promoters*, it seems more important for them to have political and cognitive resources at their disposal, than legal status and economic resources. The possibility to gain a broad consensus about the need for realising the facility, and the availability of information and scientific support about all the variables the project must take into consideration seems to be of great importance. With respect to the legal status, however, it seems that conflict is usually increased and not reduced on the basis of the institutional distribution of formal roles.

On the other hand, for the *opposers and the gatekeepers* it will be important to dispose of political and legal resources so that to co-opt as many subjects as possible in opposing the project, and to maintain their participation at a high level throughout the process as well as legal formalisation, and the possibility to access the courts, etc. seems more important than other resources.

Another relation between the role of an actor and the resources it needs to achieve success in decision making - in particular with respect to the siting of a waste facility - concerns the usage of time by the actor. The availability or lack of time cannot be intended in itself either as a resource or a constraint on success. The point is how time is used by the actors in order to achieve their objectives. In the case of waste facilities siting we could say that often the *promoters* frame the problem, highlighting the urgency of finding and implementing a solution. This perception of lack of time leads to failure in many cases, because the promoters tend to underestimate, as a waste of a scarce resource, the importance of a consensus building approach, which is always time consuming. It is very important, for this reason, when analysing a decision making process on the siting of a waste facility to verify how, in successful cases, time (in terms both of scarcity or availability) is used by the actors to drive the process toward an agreed solution or to a failure of decision making.

This leads us to discuss another aspect related to actors' resources: *the use of strategy*. Above we have identified the resources each type of actor needs to succeed in the network. If we look for success factors, in terms of the elements responsible for the positive outcome of the process, we have to concentrate on the use of strategy. In decision making processes related to the siting of waste facilities, characterised by crowded networks, a high level of conflict, an emotional approach to the issue by some of the participants, and an unbalanced distribution of costs and benefits, success largely depends on the use of strategy as a resource for problem resolution by some of the actors involved. A deadlock can often be broken by a strategic act of some actor.

Conclusions. In many respects the decision making on the siting of a waste facility is special in relation to decision making in general. This leads us to the completion of the general elements of public policy analysis for the specific field of siting waste facilities. Especially with regard to the context, the problem, the stake, the interactions within the network, and the actors' resources we can distinguish specific aspects in deciding on waste facilities siting.

Specific contextual variables that may influence decision making on waste facilities siting concern for example the social and economic situation in the region. Furthermore we can think of the degree of environmental consciousness or the occurrence of specific disastrous events.

Another important aspect in analysing cases of waste facilities siting concerns the problem definition. We suppose that a problem definition that is the outcome of an interaction including most of the actors involved, and in particular the potential opponents, facilitates successful decision making.

Related to the stake of deciding on siting a waste facility we focus on the availability of alternative solutions as well as on the presence of compensation opportunities. Both factors can contribute to successful decision making.

Specific issues on the interactions within the network that decides upon the siting of a waste facility concern the level of polarisation. In general the chances of success should be serious as the process is less polarised. Furthermore we hypothesise that a high level of polarisation can easily turn into a low level of polarisation, facilitating a successful decision making, if it is recognised during an early stage of the decision making process. Other specific issues about the interactions within the network concern the presence of a mediator and the level of formalisation of the rules of the game.

Finally we distinguish specific issues about actors' resources. The use of specific resources, and the way in which this usage can lead to successful decision making, is related to the role the actor plays within the process. Another issue about the actors' resources that needs our attention here is the strategy used by the actors involved.

1.3 SUCCESS: BACKGROUND, SHORT AND LONG TERM FACTORS

This section is dedicated to the analysis of the results of the national studies, in respect of the identification of success factors, and the indications we can draw in order to facilitate the effectiveness of public policies in waste facility siting.

Starting from an analysis of the policy problems in the seven case studies, the first part (1.3.1.) includes an ex-post discussion of the field of analysis of the research. The hypothesis is that the case studies give significant results that are not only related to waste facility siting, but more, to processes characterised by the same constraints on success. The most important implication of this conclusion is that the *indications in terms of policy redefinition are applicable to a wider range of policies.*

In a second part (1.3.2.), by analysing the structure of policy networks in the seven case studies, we try to identify the major constraints on success. Two main constraints have been identified: the crowdedness of policy networks and conflict. The main result of this analysis is that, *even in success stories, it is difficult to identify the capacity by the actors to anticipate the constraints, and conflict in particular*. Success is the result of a learning process which is part of the decision making process itself.

The third part (1.3.3.) is dedicated to the identification of success factors. They have been divided into 'background' factors, 'short term' and 'long term' factors.

Background factors have to do with the context in which the decision making process takes place: they are connected with the issue of waste facility siting, but are not generated within the process. *Short term factors* are actually the ones that are directly related to the process, and that are manipulated by the actors involved. By identifying the *long term factors* we try to draw from the analysis some indications on how waste treatment policies could be redefined in order to make them more effective.

Before going on to the analytical description of the results, it is important to recall that the objective of the research is not to *compare the performance* of seven policy networks 'at work' on the same issue. As the national studies demonstrate, the economical, social and political contexts in which the processes takes place are very different. And the extension of the project from the original five case studies to the final seven, including two Eastern European Countries, increased this diversity.

The effort here is to find the only factors that are able to explain the successful outcome of all (or, in some cases, the large majority of) the case studies. According to this point of view, the possibility of identifying factors that are verified in decision making processes taking place in different conditions reinforce the results, in terms of the possibility of drawing general lessons for the improvement of the effectiveness of public policy making.

1.3.1 If the facility is the solution, what is the problem?

A first relevant result of the empirical research is about the relation between the solution and the policy problem. The seven case studies have the final realisation of a waste facility in common: they have been selected by looking at the outcome. But, if we look at the case studies, we can easily verify that, for most of the participating actors, the facility itself is perceived as the solution to a problem that is something other than waste treatment. And consequently, that the outcome of the decision making process is perceived as successful because the implemented solution responds to that - 'different' - problem. Furthermore, this perception is not limited to one single actor (even if it is an important one): in most cases there is a 'hidden' agreement within the network around this modification. The actors, through interaction, do recognise that waste treatment is often a secondary issue or, better, that waste treatment constitutes an opportunity to play another game.

As a consequence, it is the stake that is different from waste treatment. Costs and benefits of the proposed solutions are evaluated by the actors involved, referring to a stake that is much more complex and diversified, from one case to another. And, often, some of the typical issues arising with reference to waste treatment - environmental impact, social acceptability, safety, etc. - are actually used by (some of) the actors instrumentally, in order to get to their contingent objectives. To understand the 'real' stake is consequently very important in the interpretation of each decision making process. A brief description of the seven processes could contribute to a better explanation of this question.

The *Italian case study* describes the typical situation in which the existence of a 'real' problem is taken by an actor as an opportunity for getting to its objectives. Right at the start of the process we have a (local) waste treatment problem. The first facility realised in Modena is a solution to a waste problem. But after that, the realisation of two further plants responds to the objectives of the local public agency, in terms of development of its activity. The outcome is that the agency starts 'looking for waste' somewhere else. The problem becomes the increase of market shares within a market oriented strategy by a public agency.

In the *Spanish case study* most of the actors are playing a political game. Actually this is the only process where the political parties play a significant role within the network. The project for the realisation of waste facility is taken as an opportunity by the parties - and by some individual political actors - to increase consensus at the local level. The NIMBY syndrome is used instrumentally in order to get to this objective.

In the *Swiss case study* the policy problem is directly influenced by the environmental accident that occurred in Basle just before the start of the decision making process. The realisation of the facility is taken by Ciba Geigy as an opportunity to reconstruct its 'image' in front of the other actors.

The *French case* is characterised by the relevance of the crisis of the local economy. The issue of waste facility siting is seen by all the participating actors 'through the lenses' of the impact of the proposed solutions on the occupational problem at the local level. And even if the environmental impact of the facility is extremely relevant in determining the positions of the actors involved, the proposed project is acceptable because it is perceived as not conflicting with the economical issue.

Slightly similar to the previous is the *Slovenian case*. Here, too, the occupational problems at the local level contribute significantly in determining the objectives of the participating actors. At the same time the decision to realise a new facility depends mainly on the redefinition of the market strategies by the waste producer. The problem in this case is related to the effects on the local economy of this process of organisational restructuring that is determined by national and international variables.

Apparently we have only two cases where the policy problem remains almost exclusively a problem of waste treatment: the Dutch and the Hungarian. But we can find interferences here, too.

The *Dutch case* has some similarities with the Italian one, due to the fact that the actor operating the facility is a private company, adopting a pro-development approach to the problem of waste treatment.

In the *Hungarian* case it is the dimension of the waste problem that makes it so relevant in the decision making process. On the other hand some aspects - in terms of the conflicts linked to the intervention of economic actors from abroad within the new market system - can be identified to explain the behaviour of some of the participants.

The consequences of this tendency toward the diversification of policy problems and the marginalisation of the waste problem, in terms of effectiveness of the decision making process, are not similar in the seven case studies.

We have some cases where the emergence of a problem different from waste treatment facilitates success. This is evident for example in *Salaise*, where the issue of the impact on the local economy (occupational problem) is very important in facilitating the acceptability of the new facility replacing the old, polluting one. This is actually a case where it would have been very easy to get to a 'negative' solution, simply by closing the existing plant.

Another example is the *Swiss case*, where the problem definition by Ciba Geigy - to gain confidence in front of the other actors as a non-polluting company - is an important condition for the openness of the decision making process.

On the other hand, in some cases the redefinition of the problem has negative consequences on the decision making process. The clearest example is the *Palos de la Frontera* case, where the 'politicisation' of the issue generates the most relevant constraint, increasing the conflict.

In the case of *Modena*, the adoption of a market oriented strategy by AMIU is at the same time a condition for the development of the capacity of the Centre for Waste Treatment and a factor that induces conflict.

We can now state that the complexity of these processes seems to depend in the first place on the *'overload' of different problem definitions*. On the other hand this result could take us to the conclusion that we are not talking about waste treatment policies. The following points could contribute to articulating this position, in terms of the delimitation of our field of analysis.

(a) We are not looking for success factors in waste policies. This would have caused us to consider a much wider range of solutions. It would be difficult in fact to demonstrate that realising a waste facility is the best solution to that specific problem, at least because the success (in operating terms) of a waste facility depends on the existence of waste, and more, on the increase of waste, as shown by some of the case studies. We intentionally started from the solution because of the acknowledged complexity in making this sort of decisions.

(b) On the other hand, the redefinition of the problem is - as we can find in the policy analysis literature - not a peculiarity of waste policies. The relation between a solution and the problem as defined by the participating actors is always ambiguous. To play a 'different game' is often the reason why an actor decides to join the network. More, by adopting a strategic approach, manipulating the stake (or the perception of what the stake is by some of the actors involved) can be a relevant resource to facilitate success.

(c) On the basis of the previous points, we hold that the outcome of the research is the identification of factors that are able to explain the success of decision making processes with certain characteristics (that is, with the same type of constraints). These factors, of course, are applicable (but not exclusively) to decision making processes about waste facility siting.

Our field of analysis is thus at the same time smaller than waste policies (because realising a facility for waste treatment is not the only available solution to a waste problem) and larger than waste policies (because the indications in terms of improvement of the effectiveness are applicable to other policies which are similar for what concerns the type of constraints).

1.3.2 Success through complexity

Success in these stories does not depend on the absence of constraints, even if, to a different degree, the seven processes are characterised by their complexity, and by the difficulty in reaching a positive result. This is what we expected. The aim of the research was not to find the conditions facilitating the development of 'simple' decision making processes. On the contrary, the objective was to identify the factors able to explain how success is reached 'under adversity'. This is why the case study selection has been based on indicators of complexity. But it is important to note that in most countries the selection opportunity has been limited because of the very limited number of waste facilities realised in the last decade.

Success, in waste facility siting, is linked with complexity in decision making. As a consequence, factors for success are those that are able to explain how, in these case studies, constraints have been overcome. And, to identify success factors, we have first to look at the constraints.

As with 'success', the use of the term 'constraint' in this context needs an explanation. After all, we are talking about success stories, in order to find success factors. The positive outcome of the process is to be intended as the result of all the characteristics of the process itself. It would, for example, be a value judgement to state that the crowdedness of the network is a 'negative' element. Actually, we already pointed out in the theoretical framework that some elements of complexity, that are usually intended as negative, can play a positive role in the construction of success.

So the term constraint is used here to highlight the elements of complexity that need to be approached strategically by the core actors in order to use them as resources for the solution of the policy problem. In the absence of such an approach, they could very easily turn out to be potential factors for failure.

Complexity, in the seven case studies, has to do with two main elements. The first one is the (relative) crowdedness of the policy networks. The large majority of the decision making processes are populated with many actors. This characteristic of a policy network cannot be taken, speaking in general terms, as a constraint: for example, a large number of participating actors can contribute to reduce the possibility (or the impact) of the juxtaposition between two conflicting actors, avoiding the risk of paralysis of the process. But in the processes we are analysing the presence of many actors corresponds to the presence of a broad spectrum of objectives, which are difficult to integrate around a specific solution to the policy problem. As described in the national studies, the constraint is determined by the multi-dimensional structure of the interaction among conflicting objectives. To go on with the analysis, we have then to look at the characteristics of the actors.

The complexity of the decision making processes has to do with their crowdedness not only in terms of the number of the actors involved, but even in terms of the diversity of the actors. Diversity can be read on two different dimensions.

On a vertical dimension, we have actors at different levels. Almost all the case studies see the intervention of institutional actors at the central level. But we have private actors too that, even if involved as owners of a specific plant, are actually national (or international) actors. On the other hand, all the processes are characterised by the presence of local actors (even at the neighbourhood level).

On a horizontal dimension, we have institutional actors (city councils, public agencies, departments, etc.), political actors (political parties, single political figures), social actors (local opposition groups, environmental groups, etc.), entrepreneurs (waste producers, waste 'treaters').

Within this general frame, the seven case studies are obviously quite different from one another. The less crowded network is the one activated on the *Dutch* decision making process. Even the *Italian case study* is characterised by the participation of a quite limited number of actors, mainly at the local level.

A broad spectrum of actors constitutes the network in the *French case study*, including national, regional and local actors. Confirming the expectations, institutional actors are over-represented with respect to the other networks.

A very crowded network is present in the *Spanish case* too. Here the peculiarity is represented by the significant role played by the political parties, which is quite exceptional as compared to the other case studies.

The *Swiss case* is particularly complex in respect of the composition of the network. Apart from the fact that the analysts counted about 30 participating actors, it is their diversity that makes this network very broad. Almost all the categories are represented (social, political, economical actors, at the central and local level), but more, we have an international dimension, determined by the location of the city on the border with Germany.

In some way an international dimension is present in the *Slovenian case* too. In fact, while the actors are all Slovenian, with a presence of both national and local ones, for some of them the international dimension is relevant to the definition of their objectives. Complexity depends on the vertical dimension of the network, with some actors - such as the waste producer - behaving within an international frame, and other actors looking at the very local aspects of the decision.

The complexity of the *Hungarian case* depends mainly, as pointed out by the research team, on the presence of a large number of individual actors. In the analysis of the decision making process it is difficult to identify the objectives of the representatives with the objectives of their organisations (both public or private). This probably has to do with the transition phase Hungary is currently facing.

The second important constraint to success is conflict. We can always identify - in the seven case studies - an actor playing the role of the *promoter*, with a strong commitment (responding to different objectives) to the realisation of the waste facility.

The promoter is played by different figures, mainly depending on the public or private nature of the actor responsible for waste treatment in each situation, as follows:

- the local public agency responsible for waste treatment, in conjunction with the city council (AMIU) in Modena;
- a private company (VAM company) specialised in waste treatment in Wijster;
- the regional agency responsible for waste treatment planning (Andalusian Environmental Agency) in Palos de la Frontera;
- a waste producer (Ciba Geigy) in Basle;
- another waste producer (LEK) in the Slovenian case study;
- the Mayor of Garé in the Hungarian case study;

- the French case study is more complex than the others in respect of the identification of the promoter: we have on one side the private company operating the existing waste facility (PEC-TREDI as a subsidiary of EMC) promoting the existence of a waste facility; on the other side the Mayor of Saint Maurice, promoting the solution of the pollution problems of the existing plant in Saint Maurice.

The role played by the promoter is decisive in all the case studies: by mobilising resources, by activating the other responsible actors, by favouring the interaction, the promoters represent the 'focal point' of the decision making process.

It is important to note that, with respect to other 'types' of networks, in these case studies the promoter plays a significant role throughout the process. After having raised the problem, usually by presenting a solution - the project of a waste facility - the promoter manages to conduct the process toward a positive outcome. This often depends on the fact that this role is played by the actors responsible for waste treatment and for operating the facility. What we have here is actually a figure of promoter-director: a concentration of resources for the implementation of that particular solution.

But, as we have consistent interests pressing *for* the realisation of the facility, we often have consistent actors *against* the realisation of the facility. The opposers are much more homogeneous than the promoters: they are mainly social, local actors (and sometimes their representatives, or 'supposed' representatives):

- local and environmental groups in Modena, in Slovenia, and in Wijster, all of them looking at the environmental consequences the facility;
- a more articulated group of opposers, including political parties, in Palos de la Frontera;
- an international set of actors, including local and environmental groups, city councils in Switzerland and Germany, in Basle;
- the citizens and city councils of the villages around Garé, fearing that they will pay the costs (in environmental and health terms) of

the realisation of the plant, without getting benefits (in terms of compensation);
- the operating company in the French case study, trying to avoid the costs of building a new facility substituting for the old (polluting) one (but we also have local actors opposing the new facility project).

It is fairly clear, now, that a typical constraint on success in decision making processes is not present in the seven case studies. While many processes risk failure because of the lack of interest by the responsible actors, and success depends on the capacity by the promoter to represent the solution as a way to distribute benefits, in order to mobilise the network, in the seven case studies we have the opposite. It is the concentration of conflicting interests about waste facility siting that makes it difficult to get to a positive outcome.
This concentration has two main consequences which are relevant for the analysis of success factors.
(a) As a general rule, it is very difficult to reduce the complexity of the network by trying to reduce - by use of legal resources - the number of the participating actors. These processes are very open: non-institutional actors have consistent resources which are mainly determined by the impact of the issue of waste treatment on public opinion. As we will see there are exceptions, but they tend to confirm the rule.
(b) The networks are characterised by the availability of a consistent amount of resources (economic, legal, political, cognitive). The problem is not scarcity, but the capacity to use strategically the available resources in order to construct an agreement around a solution, involving the large majority of the participating actors.

If the main problem in reaching a positive outcome is to solve conflict in crowded and heterogeneous networks, and if this problem is taken in charge by the promoter and its 'allies', the relation between conflict and the strategies enacted (for solving it) is a crucial question in understanding success.

In other words, in most cases success depends on the capacity of (some of) the actors to 'solve' the conflict. By 'solving' we do not necessarily mean that an all-inclusive agreement is finally reached, but that, if some actors continue to represent the decision as negative, they do not perceive it as convenient to continue to oppose it.

It is important to note that, if this happens in some cases (for example in the Swiss one, as regards some of the German actors; and the Wijster case, as regards the local groups), the losers are often the actors 'instrumentally' using the issue of the waste facility, being driven by different objectives.

The resolution of the conflict, going back to one of the variables considered in the theoretical framework, has to do with the change of the *stake*. What actually happens in most cases is that, starting from a zero sum game, the decision making process evolves, turning the perception among the actors into a positive sum game. At the end, the large majority of the participating actors do agree with the proposed solution, as it is eventually redefined during the process, because it distributes more benefits than costs among the participants. How this positive outcome is reached is the subject of the following chapter.

This positive evolution of the decision making processes can be verified - from a different point of view - if we look at another of the variables, that is the patterns of interaction. In general all the main patterns identified in the theoretical framework are present in the seven case studies: confrontation, bargaining, and problem solving. As conflict is the main constraint on success, we can more often find a confrontation pattern, while it is quite difficult to find a real, all-inclusive, problem solving one.[1]

It is interesting to note that at least in one case - the *French* one - the adoption at the start of the process of a confrontation pattern is actually a condition for success: the problem, in Salaise, is to exert pressure on the operating company about the polluting effects of the existing plant. But what the processes have in common is the general tendency to evolve from a confrontation pattern to a bargaining pattern, and eventually, at the end, to a problem solving one.

[1] While in most of the national studies the patterns of interaction are treated in an analytical way, taking into consideration the separate relations among different groups of actors, in this chapter, in identifying the patterns of interaction, we refer to the type of relation among all the actors involved in a specific phase of the process.

Apparently there is just one case where the persistence of a confrontation pattern does not reduce the chances of (if it is not an actual condition for) success: the *VAM Wijster* case. In this case the already extensive resources of the promoter are reinforced by the 'mistakes' of the opposer: the lateness in joining the game; the exclusive use of legal resources. And, by the way, the difficulties in opposing a project on a basis - the expected negative public health effects - that looks difficult to legitimate before the authorising authorities. But here, too, we do not have a 'pure' confrontation game. In fact, at the end of the process, some sort of compensation is included in the solution implemented.

The evolution of the decision making processes from a zero sum game to a positive sum game, from a confrontation to a bargaining pattern of interaction, is determined by the change of behaviour of at least some of the participating actors.

Success is then facilitated by the enactment of related strategies by some of the actors involved (with an important role played by the promoter-director): in all the case studies these strategies are defined within a learning process that is itself an outcome of the interaction.

In short, what we did not find is the adoption of an anticipatory approach by the promoters-directors, facing the problem of building consensus before conflict occurs, in order to strategically 'use' the presence of conflicting interests, as a resource for the construction of an acceptable solution. This is a relevant result of the empirical research, because it demonstrates how, even on an issue (waste facility siting) which is widely recognised as highly conflictual, and even in success stories, the responsible actors seems to act (at the start of the process) as if this would be a non-existent problem (until they face it).

Of course differences of approach emerge, due partly to the different degree of awareness of these problems in the seven countries, and partly to the specific objectives of the actors involved. So, in an 'ideal' continuum of awareness we could place the Swiss case at one extreme (maximum level of awareness), and the Hungarian case at the other (minimum level).

The promoter-director in the *Basle case* has an explicit attitude to the reduction of the conflict: even if the opposers tend to impose a confrontation pattern of interaction, Ciba Geigy manages to turn it into a bargaining one. This strategy is very efficient because, by accepting the demands of some of the opposers, it reduces the resources of the others (incidentally, the ones that are 'using' the issue of waste more instrumentally, to 'play another game'). The promoter here is able to anticipate the conflict, but is ready to cope with it when it occurs.

The *Dutch case* is peculiar because the somehow 'traditional' instruments of consultation work efficiently in this specific case. The promoter uses the legal resources determined by the fact that the opposers did not express their position during the consultation phase. And it works. But it is difficult to hold that it would have worked the same way with a stronger opposition. In this case, then, the promoter does not need to anticipate the conflict: the legal framework for the treatment of the conflict is sufficient to guarantee success.

Another peculiar case is the *Slovenian* one, as shown in the proposed interpretation. Here the promoter use an efficient strategy based on the attempt to *bypass* the potential conflict. The use of the term lightning rod referred to BDV, the local company treating the waste before the realisation of the new facility, is an indicator of this strategic approach by the promoter. This is the only case where, maybe - this actually remains as an open question - the strategy used by the promoter was based on an anticipatory approach. On the other hand it is the strategy itself which looks difficult to reproduce with the same level of effectiveness in a different situation (to bypass the conflict means actually to exclude potential actors from the process).

The *case of Palos de la Frontera* is interesting because success depends largely on the decision to set up the so called Clean-Up Table, a formalised arena involving almost all the participating actors, where mediation takes place. But this successful initiative (one of the most innovative approaches to the management of the decision making process in the seven case studies) occurs - as a sort of 'final solution' - after years of paralysing conflict.

Finally, two cases are less relevant, from this point of view. The *case of Modena*, because the reduction of the development strategy by the local public agency, following the emergence of the conflict at the local level, does not halt the realisation of the new plants: the problem is in part transferred to decision concerning the operation of the facility which have not been taken yet. The *case of Salaise*, because the problem, as already noted, is not to anticipate conflict, but to use as a resource in order to obtain the realisation of a new, non polluting plant.

In short, we could summarise the results of this survey as follows.

(a) The promoters usually do not adopt strategies to anticipate conflict. There seems to be a tendency to ignore potential conflicts, or maybe to over-evaluate the available resources, which gives confidence to the promoter, as if it could win a confrontation game: but this never happens. Even in the Wijster case we do not have a 'pure' confrontation game: at the end of the story, a limited compensation even takes place, as a result of the acceptance of some sort of bargaining pattern by a very strong promoter.

(b) The promoters in the seven case studies reveal the capacity to learn from the process, even if it is within the differences of awareness described before. Facing the risk of failure, they are able to define and enact appropriate strategies, and to drive the process toward a positive outcome.

(c) The resources available to the promoters are sufficient to guarantee success. But, in order to increase the chances of success in this type of decision making processes, there is a need to favour the adoption of an anticipatory approach for the reduction of the constraints on success.

1.3.3 Success factors

The core of the research, as well as the aim of the whole exercise, was of course the identification of the conditions under which the building of waste treatment facilities has been possible. One has to remember that the default situation in most instances is the failure of the proposals, due to the impossibility to overcome the local oppositions. From this point of view the general hypothesis underlying the research was to explain the successes in terms of some specific factors, while it is in general possible to explain the failures in terms of most general factors.

Following Dunn (1981), in other words, we shared a micro-positive attitude rather than a macro-negative one: in the kind of retrospective research we were engaged in, we were not looking, as scholars usually do, for the general, and someone could say generic, reasons why things went badly, but for the very specific conditions under which the policies (in our case the siting of waste treatment facilities), are able to generate the expected result (in our case the building of an actual facility).

Background factors. One of the findings of the research, which we did not fully expect, at least to the degree emerging from the evidence, is the role of some factors that can be characterised as pertaining to the environment, because they are only partly connected with the issue of waste treatment, and they are not part of the decision making processes analysed.

We referred to these factors (see Section 1.2) as the *situative variables*: those that, not being part of the decision making process (because the actors cannot modify them) can nevertheless modify the actors' objectives and/or resources, and therefore influence the final outcome. Among the examples put forward in the conceptual framework we listed the social and economic conditions (e.g. negative occupational trends in the area), the structure of institutional arrangements, the impact of other policies, the influence of specific events (e.g. accidents), the level of environmental consciousness and the technological and scientific culture.

Most of these factors are actually at work in the case studies analysed. For instance, in the Swiss case the importance of the 'historical accident' in the chemical plants in Basle cannot be underestimated. In a similar way the French case is certainly characterised by a preoccupation about the declining industrialisation of the area, and therefore with the negative occupational trends. However these elements, very useful as they are in explaining the individual instances, cannot be generalised because in most situations we can find neither previous accidents, nor a worsening of the local economic situation.

Of a more general nature, and therefore more interesting to the

conclusions of this work, are three elements that we find in many instances. For lack of a better definition we gave them the following names:
(a) the level of industrial culture;
(b) the dominance of a political pro-development policy orientation; and
(c) the level of institutionalisation of policy making.

The first factor is quite straightforward: the acceptability of waste treatment facilities is vastly facilitated by the fact that the communities involved have a tradition of industrialisation that makes them familiar with the world of modern production and, in general with the complexities of technological change. The oppositions in some cases not only came from the non-industrial groups (e.g. farmers in the Netherlands), but in general the impact of some specific factors in explaining successes (like the use of best available technology) were to have been much lower if the possibility of discussing the technical aspects would have been restricted by the low level of knowledge among the local population.

In more general terms one could argue that, because the whole issue of toxic waste is somewhat linked to industrial production, a high level of industrial culture allows the conceptualisation of the treatment plant as an extension of the production process, even if, as in most cases (but not in Switzerland and Slovenia, however), the new plant is not linked to specific existing industries.

We will return to this point. For the time being, the negative lesson we can draw from this first, not fully unexpected, point is that the efforts to locate the plants in regions devoid of industrial tradition can make the decision a very difficult one to reach, not least because the public debate runs the risk of being very 'unframed', emotional and characterised by fear.

But the existence of the industrial culture is only one part of the solution. Even more important, perhaps, is a second, related, factor, which is the dominance of what we have decided to call, for lack of a better term, a political pro-development policy orientation. The meaning of this definition is fairly self evident: all things being equal, the establishment of a new waste treatment facility seems to be easier if the community (or communities) concerned share the idea that the economic development - as opposed to the preservation of the existing level of wealth - is an important and legitimate political goal. In well developed (someone could say over-developed) Western Europe this is a condition that is certainly more difficult to find than the previous one. And, even more important, the combination of the two can actually be rather infrequent. After all one could expect to find a clearer pro-development political orientation in the less industrialised areas than in the traditional ones, while on the other hand the reverse is true as far as industrial culture is concerned.

This can be the real reason why it is often considered interesting to look for locations in de-industrialising areas, with severe occupational problems. These locations are in fact both characterised by an industrial culture and a somewhat obligatory pro-development (or anti-decline) political culture. But if our analysis is right, the link between the trend of level of unemployment and the social acceptability of waste facilities is a sort of a spurious correlation. What matters is the coexistence of industrial culture and pro-development political orientation, and this can also be found in wealthier areas, as is the case in Modena and Basle.

The combination of these two factors can therefore explain the somewhat striking over-representation of communities with a left wing political orientation in our collection of case studies. After all socialism and communism - with their emphasis on the crucial importance of industrial workers - are more likely to establish their roots in communities where industries have existed for a long time (and therefore have an industrial culture), and to share the belief that more development, and more industries in particular, are a good thing.

The third important factor in this class of non-process related elements is quite different from the previous ones and refers in general to what we could call the policy making style of a given country (or, albeit less likely, of a given region). It is important to note that we believe that, even if our observations refer mainly to the waste treatment policy, this specific feature is of a somewhat more general nature and characterises policy making and implementation in general.

The critical issue here is the level of institutionalisation, that is the level to which the path to be followed in order to generate outcomes is fixed in advance, and therefore in any specific instance there is a more or less well ordained sequence of decisions, controls and procedures to be complied with.

The main finding of this research in this respect is that a *certain* level of institutionalisation of policy making and implementation is an important situative variable in explaining successful outcomes. In practice this means that a situation in which no procedure whatsoever is defined for dealing with the problem of siting a waste treatment facility, or, on the contrary, the existing procedure is too strict in legal and/or bureaucratic terms, seems to be a very unlikely candidate for a successful outcome.

A 'good' procedure, or the 'right' level of institutionalisation, is a situation in which, on the one hand, there is an established way of dealing with the question at hand, and on the other hand the legal rules are flexible enough to provide a sufficient leeway in order to adjust the elements that are peculiar to each different decision making process.

When we discuss the process related factors explaining the individual successes we will discover the importance of this feature. One cannot deny, however, that this variable is probably more related to the overall working of the political and administrative system than with the issue of waste treatment policy. The required level of institutionalisation is the one reached by those countries that seem able to avoid the opposite dangers of too much or too little reliance on meta-decision making.

From the point of view of the creation of the legal instruments for waste treatment policy, we could say that the total absence of any provision to secure an at least partly planned and co-ordinated process is at least as dangerous as the idea that the planning exercise should define every single step with extreme precision. To the waste industry we might suggest that to define the rights of public enquiry and consultation seems to be an effective way of eventually bringing about a decision. To the opponents of waste facilities we might suggest that the emphasis on legalistic aspects may finally be counterproductive in avoiding 'bad' solutions.

But there are limits to institutional engineering, of course, and it is possible, indeed likely, that the level of *de facto* institutionalisation of a policy is dependent on the constitutional rules governing a country, and therefore the way in which the processes are managed is bounded by this background factor, or the policy style of a given country.

Short term factors. But of course the central point of our research was to look for the very specific factors, directly related to the processes analysed, that are potentially and actually manipulated by the core actors of the decision making process. We have tried to put forward and argue some hypotheses about those factors. Some of them were strongly confirmed.

As is clear from the analysis of the case studies, a certain amount of process management can quite invariably be found. This, in some instances, amounts to fully fledged strategies on the part of the promoters of the facility (one could think again of the Swiss case), while in the other, possibly the majority of cases, there are only fragments of strategy, and more often than not the success is also the result of non-intentional, contingent, factors (in general all the cases where the success is closely linked with a previous failure, as in Spain). But some strategy (that is a certain amount of usable knowledge about the process), seems to be, in accordance with what we expected, a crucial element for the success of the decision making process.

Implicit in this first finding is the fact that the role of the promoter can also invariably be found, as the one actor who should develop the strategy and try to manipulate the crucial elements of the process. What is interesting is that the nature of this promoter seems rather irrelevant: we find private companies (as in Slovenia and Switzerland), local institutions (as in France or Italy), and political parties (as in Spain).

The form of the networks tends to vary from one case to the other. The only generalisation that seems possible in this respect is the fact that sooner or later, for a short period or for a longer one, the critical character of the issue generates a situation of total interaction, as defined in the research framework. But this element does not seem to be associated with success in a meaningful way.

Apart from these quite general findings, the more interesting results of the analysis can be regrouped, again following the scheme of the conceptual framework, in the following way:
(a) factors concerning the problem;
(b) factors concerning the stake;
(c) factors concerning the rules of the game.
It is worth noting that some other groups of factors (the already mentioned structure of the policy network, or the actor resources) seem to be of lesser importance in explaining the research results.

We have already shown how it is possible to argue that in most - and maybe all - cases, the *real* problem is not actually about waste. We will not come back to this point, but, among the specific, or short term, factors explaining the successes in siting a waste treatment facility, we can certainly list the fact that the chances of this outcome are higher if the actors share the idea that the problem is not one of waste disposal, but one of waste production.

Put in another way: the success will be easier if the actors concerned share the belief that they have a waste problem to solve, where the emphasis is on the fact that it is their problem. This is certainly the case in Hungary, in France, in Switzerland and in Slovenia, but it is also the case in Italy and Spain (where the definition of the geographical area is somewhat larger).

This is certainly a point to bear in mind when planning new plants: all things being equal the chances of success are much higher if the problem of waste production, and therefore disposal, is already on the agenda. The most interesting paradox in this respect is the one provided by the *French case*, where, quite apart from the existence of chemical works, it is the very existence of a previous, polluting treatment plant that actually seems to help, rather than hinder, the establishment of a new one. But also the *Swiss case*, where the local population well remembered the accident of some years ago, seems to prove the point we have made, and the same is true about the *Netherlands case*, where the existence of a waste treatment tradition is considered part of the explanation.

But the acceptance of the plant as the solution to the already existing local waste problem is not enough. In an ideal checklist of the factors explaining the success, the issue of which kind of plant seems very important indeed.

One aspect that cannot be underestimated is the crucial importance of stressing that the facility will be run with the *best available technology*. Quite understandably this is a point observed everywhere, one that explains the large amount of cross-fertilisation that we find between the cases under investigation. This too could have been expected - after all, there are not many instances of new facilities that have been decided in the last ten years in Europe - but it is nevertheless striking, at least in Southern and Eastern Europe, where the reliance on the domestic technological capabilities seems to be lower, that the use of foreign know-how seems to be a guarantee that the best available technology is actually employed. So the Italians went to Belgium in order to find the inerting process, and the Spaniards adopted the same process, after visiting the Italian plant. The Hungarians made an international tender that was decided after a visit to the French plant.

Furthermore, in the cases in which the solution is domestic, or at least the fact that the same process is used abroad does not seem to be an important argument, the stress on the technological novelty of the solution adopted is very important in the communication strategy.

But there are other aspects worth noting. The first is that a supplementary guarantee against a high level of nuisance to the local population is represented by the fact that the proposed locations are very often within an industrial estate, as for instance in *Spain*, in *Slovenia* and in *France*. This too seems to be quite an obvious solution, but in fact it does not mean that it is less important, at least because this often implies that it is possible to locate the facility not too far away from the residents, provided it is in the industrial area: this is certainly the case in *Basle* and *Modena*, but may be true in other cases, too.

All in all we can say that a *careful stake management* is a necessary ingredient of the recipe for success. Apart from the already mentioned stress on best available technology, and location within existing industrial estates, we should obviously refer to the question of the need to clarify very well from where waste - mainly if it is toxic - will come. This is partly connected with the point made above about the need to link the proposed solution with a local problem, but it is more general in nature.

The promoters of the creation of a treatment plant must understand why the local population and the opposers want to know the origins of the waste, and why the provision of such knowledge is in the interests of the promoters themselves. The first crucial point, here, is the *level of uncertainty about the effects of the plant*: any element able to reduce this uncertainty is positive insofar as it prevents the spreading of uncontrolled fears.

But of course this is not merely a symbolic aspect. The second aspect is that to define the geographical area generating the waste and/or the type of waste to be treated in the proposed plant is very important because it is one of the negotiable aspects. Because in the course of the process it is possible to change the one or the other, and to redefine the project accordingly.

Stake management, in this form, means, therefore, the attempt to find the negotiable issues, and the readiness to alter the project in order to adjust the different positions of the actors. From this point of view it is equally wrong - or at least dangerous from the point of view of the success of the individual processes - both to take a rigid position on the need to leave the facility totally free in getting all the waste they think they are able to treat, and to imagine that the issue can be dealt with once and for all through a planning exercise, without any relations to the specific siting decision. After all, the Slovenian and the Italian cases show how one of the most critical factors has been the unilateral decision on the part of the managers of the plants to take waste from 'abroad' without any sort of public consultation.

But stake management - as part of process management and strategy building - means any change in the content of the decision, even in a merely symbolic way, able to adjust the interests of the actors.

From this point of view the ability to make *package deals*, linking the building permit for the proposed plant with some other issue of public or private interest, is another point of interest. For instance, in the *Netherlands case*, the creation of a small fund, financed by the waste company with one guilder for every ton of treated waste, aimed at promoting waste reduction, is little more than a symbolic gesture, but can help in the negotiation process.

And *negotiation* and *bargaining* are definitely the key words as far as stake management is concerned. This can go as far as the Swiss case, where one can say that the project was actually formed out of the results of the consultation phase, or it can be more limited in scope. But in any case the 'take it or leave it' strategy, even if accompanied by generous compensation, high level scientific guarantees, or a large deployment of public relations exercises, is almost invariably a recipe for failure.

The third and last group of factors able to predict the successes in this kind of game pertains to what we can call *procedural management*, that is the ability to control some aspects of the process through which the decision making takes place.

We have in fact referred to this kind of problem when dealing with the more general issue of the level of institutionalisation of the policy, and both the short term factors we are able to identify are congruent with this more general orientation.

The first one we can label as the ability to *control the openness of the network*. Here what matters is the attempt to avoid the opposite dangers of polarisation, brought about by too few actors, and of a never-ending and cyclical debate, generated by too large a network. The possibility to introduce new actors, or, to a lesser degree, to exclude old ones, is therefore extremely important for developing a successful strategy, and in any case the simplistic equations about the network of the kind 'the smaller the better' or 'the larger the better' are equally false.

This, for instance, is a warning against the dangers of generic public relations exercises, because with their emphasis on general communication to the public they run the risk of raising the interest of actors that are simply too many to be able to sit at the bargaining table. A waste facility cannot, and does not need to, be 'sold' as a commodity, and advertising of any sort can be counterproductive.

From this point of view the process should be carefully managed to include only these actors that have a stake in the issue. For instance, in the *Hungarian case* the success seems to depend also on the ability to exclude the 'outer' municipalities from the consultation process, i.e. those which are not really affected by the negative externalities of the plant, and that a too generic planning had previously included in the process.

But the reverse is also true. The dream of many private promoters, i.e. to have a straightforward process in which they apply to one and only one public administration, provide the relevant information about the safety of the plant, wait for a few weeks and get the needed permit, is very likely to become a nightmare. This approach, with no room for lateral deals and negotiations, is the most fragile. Consider only the following: if the public administration is a regional or national one, the opposition of the local municipality is a virtual certainty. If, on the other hand, the local administration is the key institutional actor, the temptation to externalise at least part of the costs (for instance by finding a location very near to the municipal borders) is so strong that the opposition of the surrounding populations is highly probable. And in any case the issues involved in our research problem are multi-functional. They include considerations about the environment, the health of the population, the economic situation, etc. To think that the network dealing with all these problems can be a very simple one is a very dangerous illusion indeed.

Even more important, in most cases one should pay attention to the fact that different categories of actors are represented in the decision making process, i.e. the scientific community, the business community, the representatives of grass roots organisations, the political actors, and so on. To have different point of views is more often than not an advantage, for the simple reason that, in case of conflict, they can provide the needed redefinition of the problem, in terms that make it acceptable to the opponents.

The conclusion that can be drawn is that the management of the decision making process, through careful and selective distribution of the information available, as is shown in the cases of *France* and *Slovenia*, is actually aimed at governing the network of actors, both through the enlargement and the reduction of the participants, as the case may be.

The second and last point concerning the rules of the game is also connected with the need for *flexible institutionalisation*, but in this case what comes into consideration is not the structure of the interaction, but rather the way in which it takes place, in two words the *procedural rules*.

Now, it is fairly obvious, and in any case could have been expected, that the patterns of interaction found in the processes range, at least in the final phases, between a problem solving attitude to a bargaining one. The confrontations that can occasionally be found in the case studies were in general solved well before the taking of the actual decision. What is more interesting is one less expected feature that we can call the widespread use of *informal agreements*. By this term we mean the existence of a negotiation, between public bodies and/or private firms, that usually goes beyond the strict provisions of the law, in some cases setting stricter standards than are legally prescribed, in others defining rights to compensation by the affected interests.

On the one hand this is another instance of *stake management*, as these deals show the willingness to enlarge the issues under discussion. On the other hand it is a cue of the flexibility of the institutional setting, and of a non-legalistic approach, even in countries - like France or Italy - where the usual operation of the public administration is bound by inflexible rules.

The most evident examples of the relevance of informal agreements can be taken from the *Spanish and the French cases*. The solution of the highly conflictual situation in the Palos de la Frontera case is the constitution of the so-called *Clean Up Table*, which becomes the arena where mediation takes place, successfully.

In the Salaise case the solution of the conflict is linked to the signature of a *protocol* by the different actors involved, which - as highlighted in the national study - 'does not have any legal value as such; but it has a very important symbolic value'.

The importance of these agreements cannot be underestimated. Quite apart from their actual content, they are further evidence of the willingness to adopt a problem solving attitude, and for this reason only, they seem to represent - at least from the symbolic point of view - a very important step in the decision making process.

Long term factors. We have now reached the conclusion of our research and we could very well stop here. Some lessons have been drawn from the evidence collected during the research, and some suggestions for future policy makers and planners have been proposed. But we feel that it is possible to go a little further and that we can try to speculate about the *optimum setting for waste policy*, or at least to make some suggestions for the time, which will come sooner or later, when the issue will be treated - as seldom happens today - in a mature and less emotional way.

The key point of our proposal is that, if there were to be a widespread acceptance of the simple truth that waste is a fact of life and not a failure or a shame, the need will arise to find the best possible balance between the role of the private sector and the role of public administration.

The present situation, at least in most countries, seems to be too heavily unbalanced toward the public sector, which often cumulates the potentially conflicting roles of scientific and technical guarantee, regulator, financier and provider of the service.

In order to achieve a better balance, and to design the waste policies for the next century, the point of departure is the *polluter pays principle*. A full adoption of this orientation is at the same time the best guarantee against the dangers of waste overproduction, and the best incentive towards an overall reduction.

The fact that the producers pay the full price for a safe and effective disposal of their waste is, in a market economy, a powerful incentive towards the change of the production process in order to avoid unnecessary costs. This is obviously more true in the case of industrial firms - and toxic waste is usually the by-product of industry - which have to face price competition. Here the development of waste-free production processes can be a very powerful competitive advantage both within the industrialised countries and vis-à-vis of the developing countries, where it is likely that, in the foreseeable future, the stringency of the environmental regulations, and therefore the price of waste disposal, will be somewhat lower.

The second point to bear in mind is that - regardless of public or private ownership - *the waste industry is a normal industry*, where the importance of technology, the need for good research and development, the weight of fixed capital, all emphasise the scope of scale economies.

The consequence is that the waste treatment industry must be treated as a perfectly normal industry. Of course it will be regulated, not least because of the potential danger to the environment, but not in a different way from any potentially dangerous industry. But this regulation should not prevent the development of the industry in an efficient and economic way. This implies that any restriction on competition (for instance the licensing system) should be regarded only as a necessary evil (and may be a transitional one), and an even more dangerous step could be the setting of a principle saying that every producer should dispose of its own waste. Apart from the technological implications of such a situation, this could effectively prevent the development of the waste industry in an economically viable way.

The third point to bear in mind is that, as we have already suggested, the overload of functions on the public sector can generate a difficult situation to cope with, and therefore *the scope of public intervention should be* somewhat *reduced* in the future.

Today in some countries the State, locally or centrally, at the same time provides the technical and scientific guarantee, the regulations, the finance and sometimes even the actual running of the service. Now, if it can be argued that the public ownership of the treatment plant - the idea that nobody makes money out of treating waste - can be considered, as it is spelled out in the *Spanish case*, a short term factor able to increase the social acceptance of the plant itself, it is also true that it is very dangerous, because it creates the conditions for underpricing the service. Not only will there be a huge investment without the compelling need of a return on capital invested, but it is also difficult to imagine that the public administration will always be able to resist the temptation to use the public ownership of the treatment facility in a protectionist way; for instance in subsidising ailing and polluting industries, with the result of postponing the changes in the production processes that would be needed, the industrial firm having to pay the full market price for waste disposal.

But this is not the only possible danger of overload. Being at the same time a producer of the service and a regulator, furthermore, the State runs the risk of using its authoritative powers in order to protect its capital investments, even to the detriment of the environment. For instance, setting the standards for the waste industry at a level that can be easily met by the existing public facilities in order to avoid expensive refurbishments.

Finally, and less obviously, there can be a certain degree of contradiction and even of conflict between the inherent political and, one might be tempted to say, partisan nature of public administrations in a democratic society and the need for a *super partes* attitude implicit in the role of scientific and technical guarantee. In a situation in which the environmentalist movement has entered the political arena, it seems increasingly impossible to cumulate the role of regulator and that of the actor who guarantees the fact that the best available technology is employed, that the projects are feasible, that the level of pollution and/or nuisances will be within the acceptable limits, and so on. The 'trust in the State', sad as it is, has already declined in modern, complex societies. To overload the administration with many different and potentially conflicting functions can be a very dangerous strategy, and eventually a self defeating one.

Of course there are at least *two reasons for keeping a wide scope for public intervention,* to the point of recognising the need for public ownership of waste treatment plants.

The first is the *lack of a sufficient level of specialised knowledge and know-how in the private sector*. The need for public intervention is argued on the basis of the costs of developing this know-how in a given national setting. But in reality this argument, with the growing globalisation of the economy, and with the abolition of trade barriers, does not cut a lot of ice.

More serious is the second argument, the one that emphasises the *need to prevent illegal behaviour*. Waste treatment and disposal can be, and in some instance already is, quite literally a 'dirty business', because there is an inherent incentive to cheat on the regulations both on the part of the waste producers and of the service producers. If the costs of illegally disposing of the waste are, as they very well can be, astronomically lower than the costs of safely disposing of it, then the dangers of collusive behaviour are very high indeed. This is one of the reasons why in some countries - like the U.S. - organised crime is heavily involved in the waste industry, the payoffs being very high and the need for secrecy being met by the existing organisational structure.

This is indeed a point to be taken seriously, even if the dangers of such unpleasant developments are not evenly distributed between the different countries and between the different types of waste. Even if we abandon the arguments about the level of civic culture in the different countries, the question here is the level of concentration of the waste industry (which can be very high, as the French case shows) and the related importance of capital investments in the industry itself. From this point of view the real danger of illegal behaviour is higher when the capital investment is lower, as in the case of domestic waste treatment and/or waste transportation. After all, a multinational firm investing very large amounts of money in its facilities is a less likely candidate for Mafia infiltration than not only a small, local firm, but also a local community owning a treatment plant. Even from this point of view, therefore, there is an argument in favour of the development of a competitive - even if oligopolistic - modern waste industry.

On the basis of these considerations - the need to fully implement the polluter pays principle, the rationale of considering the waste industry as a perfectly normal activity, and the dangers of an overloaded public sector - we can therefore suggest a *possible optimum setting* for the problem at hand.

The first point is that the task of regulating waste production as well as waste treatment and disposal should stay with public administration, namely with the technical components of environmental policy. This is a fairly obvious point but it must be emphasised. There are good arguments both for considering the formulation of the regulations a legitimate object of political debate, through the process of democratic law making, and for considering their day to day implementation as a technical task, possibly dealt with by a plurality of specialised agencies, like the health services, the territorial planning authorities, and the environmental agencies.

The second point is that there is room for a total privatisation of the waste treatment industry, which implies not only the demise of the State or Municipality-owned plants, but also a repeal of the existing barriers to full competition, including the licensing system. The latter suggestion, obviously, applies only to what is usually called ‘business regulation’, and not to the environmental or safety regulations, but in this case exactly in the same way in which any potentially dangerous industry is regulated. But this implies that even the limitation of competition implicit in the pre-determination of the area serviced by every plant must be regarded with suspicion. As we have observed in the previous paragraph, it is indeed possible that this limitation is a condition for the social and political acceptance of the facility, but this is and should be the result of the interaction between the actors concerned, not a precondition for the whole process.

A related point concerns the transitional phase. It is indeed possible that, in order to induce better behaviour on the part of the waste producers some sort of financial incentive should be devised. Apart from the questions regarding the safeguard of free trade implicit in the competition policy of the European Union, the point is that, on one hand, such policy must be temporary, and, on the other hand, that to subsidise the demand of waste treatment (e.g. a tax rebate for every ton of toxic waste certified as safely disposed of) is better than to subsidise the supply of the service (through systematically underpricing).

The third and final point regards the role of scientific and technical guarantees. Here there is an argument in favour of some sort of *independent authority*, possibly linked to the international scientific community, unburdening the public bodies from this task. Both during the decision making process about a new plant, and in the actual management of the plant itself, there is the room - in many cases the need - for such an institution in order to avoid unnecessary political and administrative conflict about scientific or technical judgements.

REFERENCES

Dunn, W., *Public Policy Analysis. An Introduction*, Englewood Cliffs: Prentice-Hall, 1981

2 REPLACING OLD FOR NEW: LESSONS FROM A FRENCH CASE STUDY

CORINNE LARRUE, JEAN MARC DZIEDZICKI

2.1 THE PROBLEM: THE TREDI-SALAISE CASE

The French case presented below is related to the TREDI-Salaise plant. This plant was built at Salaise in 1985, by TREDI, a company specialised in the incineration of toxic wastes and a subsidiary of a public company called EMC (Entreprise Minière et Chimique).[1] The TREDI-Salaise plant was built to replace an old waste incinerator, owned by TREDI and located at Saint-Maurice-l'Exil, 7 km from Salaise.

Salaise and Saint-Maurice-l'Exil are two small villages with about (respectively) 3,500 and 5,000 inhabitants, located to the south of the metropolis of Lyon. The population of these two villages belongs mainly to the working class: most of their inhabitants work in the chemical industries, inasmuch as these villages belong to the 'chemical corridor' of the region 'Rhône Alpes'. They have a strong working class culture, which explains the majority of left wing voters. As a matter of fact, the mayors of these two communes are members of the Communist or Socialist Parties.

The TREDI-Salaise plant is located in an industrial area, near a chemical plant belonging to Rhône Poulenc - of the most important chemist industrialist of France. Today, the factory employs about 90 staff, and processes three type of wastes:

- municipal wastes, about 10,000 tons/year: these wastes are mixed with industrial wastes before being incinerated;
- non-chlorinated waste, 40,000 tons/year;
- chlorinated wastes, 50,000 tons/year. This last type of waste has been incinerated by the factory since 1991, when the extension of the

[1] EMC (Entreprise Minière et Chimique), is a nationalised company specialised in the production of potash fertilisers, chemical products, animal foods, and engineering services, i.e. engineering in the field of fertiliser production facilities, nuclear and industrial waste treatment facilities. Within the engineering services branch of EMC, the industrial waste treatment facilities remained a minority activity until 1984-85. Since then, the importance of that activity has increased and PEC-TREDI now owns four important industrial waste treatment plants specialised in the incineration of toxic industrial wastes. Today, TREDI is among the two leaders in France in the field of industrial wastes treatment.

B. Dente et al. (eds.), The Waste and the Backyard, 47–67.

factory led to the building of an adequate installation for the incineration of chlorinated wastes.

The problem to be solved by the TREDI-Salaise decision making process was related to an obvious air pollution already emitted by the St Maurice waste incinerator plant. The population of St Maurice tried (and succeeded) in having this polluting plant close down. But, for many reasons explained below, it was decided to replace the St Maurice plant by a new, cleaner one, and to locate it not too far from St Maurice. The TREDI-Salaise story is a success story because this plant, currently in operation, was accepted by the population of St Maurice and of Salaise as well as by the local governments concerned.

2.2 THE CHRONOLOGY OF THE TREDI-SALAISE DECISION MAKING PROCESS

The chronology of the TREDI-Salaise decision making process can be divided into four main phases.

- *Local opposition to an incinerator in St Maurice (1976-81)*

The starting point of the story is the creation by PEC Engineering (which was later to become TREDI) at St Maurice-l'Exil of a domestic waste treatment facility with a capacity of about 15,000 t/year. This creation was proposed to the mayor of St Maurice and built in a new industrial estate of the commune in 1976. But, starting from 1977 until 1981, a growing local opposition to the St Maurice plant occurred in the face of the (mainly air) pollution emitted by the facility. The opposition originated with the mayor and the city council of St Maurice, who tried unsuccessfully to use administrative and legal means to reduce the pollution emitted by the waste treatment facility. The opposition then carried over to the population of St Maurice, which started some more visible action such as blockading the plant entrance.

More precisely, the opposition movement was developed as follows: first of all, following several complaints from the mayor and inhabitants of St Maurice-L'Exil, related to the pollution emitted by the PEC plant, the local state administration (DRIR, Regional Agency for Industry and Research[2]) imposed doubly severe regulations on PEC. But each time the pollution again became evident, after slight improvements. Then, in 1978, the city council of St Maurice-L'Exil decided to close the polluting plant, and enacted a municipal decree which was considered illegal by the Préfet[3]. Nevertheless, the Préfet compelled PEC to improve its facility, and as a consequence, the pollution emitted by the plant was reduced for several months. But, by mid-1979, facing new emission of pollution, the mayor of St Maurice-L'Exil asked the inhabitants to help him in order to obtain the closure of the plant. He succeeded, with his council, in mobilising the population, which blockaded the entrance of the PEC facility for one month. The population created an 'NGO against the PEC facility's pollution'. The Préfet convinced this NGO to stop the blockade by giving his guarantee that the plant would not pollute any more in the future. From 1979 until 1981, problems of pollution again become casually visible. Despite new guarantees given by the Departmental administration, the city council and the population of St Maurice-L'Exil announced their wish to have their village freed of the waste treatment facility. However, while the departmental state administration and PEC together improved the incineration system step by step, the opponents tried to get as many people as possible involved in their fight. The departmental administration, PEC, the local mayors and the representatives of the NGO met several times during this period. Despite their opposition, the NGO representatives agreed to let the waste facility run for a while because no other realist solution could be found; nevertheless everybody agree that the plant must be improved. The PEC manager proposed to add an expensive air filter on the stack, but only if he obtained the guarantee that the plant would remain at St Maurice-L'Exil. This conflicting period reached its climax in 1981, when opponent NGO blockaded the PEC facility entrance for 4 months, with

---

[2] The DRIR, Regional Agency for Industry and Research (Direction Régionale de l'Industrie et de la Recherche), is the regional agency of the Ministry of Industry; it is in charge of authorised facilities' pollution control, including waste treatment plants. This service implements environmental laws, on behalf of the Ministry of the Environment.

[3] The Préfet is the representative of the central government at the level of each department (France has 100 departments). The Préfet is the hierarchical head of the state administration represented at the level of departments. Therefore he formally enacts the authorisation for pollutant plants, taking into account the recommendations stemming from the regional department for Industry, Research and Environment (DRIRE - Direction Régionale de l'Industrie et de la Recherche et de l'Environment -) which handles technical supervision of the plant. The Préfet is responsible for public safety within his department. And a Sous-Préfet represents the Préfet at a sub-departmental level. We usually use the term of 'departmental' or 'prefectoral' in this text to qualify the administration which is under the authority of the Préfet.

the support of most of the local inhabitants and some local mayors. Several meetings were organised, but without any results. Opponents insisted on the closing of the PEC plant while others actors - i.e. PEC and the departmental administration insisted on the continuation of the toxic waste facility. The Ministry for the Environment then intervened twice in order to resolve the conflict; the central administration agreed to transfer the plant. Each actor, including local mayors and PEC workers - tried to get the NGO to end the blockade.

- *... which led to negotiating the transfer of the plant (1982)*

The NGO ceased its action when it reached the agreement from PEC and administration to transfer the polluting plant somewhere else, within a two year period. A protocol was signed by the Ministry of the Environment, the local administration, PEC, two local mayors and the NGO. This stated the rules for the transfer of the St Maurice plant to a new location, and for the improvement of air pollution emission at St Maurice in the interim. As a consequence, the incineration system is equipped with an expensive filter facility, able to abate the air pollution emitted by the waste treatment plant until its transfer to a new location.

- *A search for a new location for the plant (1982-84)*

The next period was devoted to the search for a new location and to the setting up of technical and financial plans for the transfer of the plant. By the beginning of 1983, the local mayors found a new site for the transfer of the toxic waste facility, and in may 1983, the city council of Salaise-sur-Sanne - a village located 7 km from St Maurice-L'Exil - accepted the new TREDI plant (PEC became TREDI at the beginning of 1983). The new site is located in an industrial estate. This verbal agreement had to be formally accepted by the local population. The industrial estate where the new plant was to be sited is devoted to the chemical industries and is managed by a 21 village union (including St Maurice-L'Exil and Salaise): the 'SIVOM de Roussillon'. In June 1983, the 'SIVOM de Roussillon' finally agreed with the city council of Salaise to welcome the TREDI plant.

In November 1983, a commission created to follow the transfer of the plant met for the first time. It comprised the city council of Salaise, the local mayors, the 'A.V.R.' - an important NGO of mayors who are active in the safeguarding of the Rhône valley environment, some representatives of the 'SIVOM de Roussillon' and some TREDI and

Rhône-Poulenc workers, the local technical administration responsible for pollution control, and TREDI management. Meanwhile TREDI sent to the administration its technical, formal request for administrative authorisations. But in April 1984, in order to gain one more years delay, TREDI decided to close its St Maurice plant and to dismiss its workers in June 1984, arguing the insufficient guarantees about the implementation of its Salaise project. Moreover the investment was still not entirely funded, which explains the need for a delay. Therefore, the city council of Salaise decided to accept officially the building of the toxic and domestic waste facility by TREDI. The public consultation had not yet been controversial and the local population seemed to agree with this decision. And, in June 1984, despite its initial opposition, the 'NGO for the defence against the PEC facility's pollution' decided to accept the one more years delay until June 1985. Consequently, TREDI withdrew its decision to dismiss its workers.

- *Implementing the transfer of St Maurice plant, running and expanding the facility (1985-1991)*

The last phase is related to the settlement, the operation and the extension of the new Salaise plant, which was opened in 1985 by the Ministry for the Environment. This facility is able to incinerate 40,000 tons of non chlorinated toxic waste and 10,000 tons of domestic waste per year. A control committee composed of representatives of local governments, AVR, departmental state administrations and the plant manager meets once a year in order to follow the plants activities. Finally, in 1991, the TREDI plant was extended in order to be able to incinerate chlorinated wastes, despite the initial TREDI promise never to extend its plant. Then, today, the new capacity of the TREDI-Salaise plant is about 100,000 tons per year and the plant employs about 90 workers. Technical guarantees are given to the mayor of Salaise by the manager of the plant.

2.3 THE MAIN ACTORS INVOLVED, THEIR STRATEGY AND THE INTERACTION PATTERNS DURING THE DECISION MAKING PROCESS

Among others, four main actors have played an important role during the decision making process: the PEC-TREDI team at the local and the national level, the central and departmental administration, the local (mayors) and departmental elected representatives, and the mobilised population, represented by the NGO 'against the PEC facility's pollution'. Table 1 summarises the respective levels of intervention of each actor.

Actors / Level	Company	State Administration	Elected representatives	Population
National level	• EMC (Entreprise Minière et Chimique) management board • PEC-TREDI central management board	• Ministry of the Environment		
Regional level		• DRIR (Regional Agency for Industry and Research)		
Départemen-tal level		• Préfet and sous-Préfet • DDASS (Provincial Agency for Health and Social Affairs)	• Départemental Assembly representative (Conseiller Général)	
Local level	• Manager of TREDI-Salaise plant • TREDI-Salaise workers		• Mayor of Saint Maurice • Mayor of Salaise	• Population of Saint-Maurice

TABLE 2.1: Overview of the main actors involved within the decision making process

For each of the four chronological phases stated above, we will present the network of participating actors and the interaction pattern among these actors.

Phase 1 (1976-1981): Local opposition to an incinerator in St Maurice
Before the beginning of this period, when the St Maurice plant was built (1973-76), a technical arena prevailed, dominated by two main actors (the PEC management board and the departmental state administration) which decided on technical grounds on a closed decision making process. But from 1976 and during all the conflicting phase, this technical arena would have many difficulties in maintaining itself as such. First a political platform, then a socio-political one (including the representatives of inhabitants) intervened and tried to break out of the closed technical decision process. Therefore, the negotiation process became increasingly complex and open, ending in a major conflict.
The three main arenas involved are the following (cf. Chart 1):

1. The first one is the social arena, represented on one hand by the NGO 'against the PEC facility's pollution' and on the other by the PEC management team and the NGO created by the plant workers. The opposition between these two main opponent actors impacts this conflictual period. The NGO 'against the PEC facility's pollution' claims and acts in order to be recognised as a participating actor, and to be recognised as the representative of the St Maurice inhabitants' interests. During this phase, this NGO has played the most important opposition role. As they did not succeed by 'soft' means (dialogue, petition, etc.), they implemented 'hard' means of opposition (twice blockading the entrance to the plant in 1979 and 1981). For them, the problem is mainly a problem of nuisance stemming from the waste treatment plant. So their main objective is the improvement, then the closure of the facility. This objective, and the fact that their previous attempt at dialogue were ignored by other actors, let them adopt an inflexible position. During this phase they were able to mobilise most of the St Maurice people and to use political resources by means of support from elected representatives at the local, regional and also national level. The stake for them was to obtain the right to be involved in the decision making process. They intervened on a confrontational basis, each time trying to break the 'natural', closed technical arena (which involved mainly the DRIR and the company).

The PEC management team helped by the NGO created by the plant workers, which wanted to counterbalance the opposition stemming from the population, constitutes the second side of the conflict. During this phase, the company considered the problem mainly as an economic one. The main objective of the company was at that time to preserve the waste treatment activity at St Maurice. In order to reach this objective the company mainly mobilised economic resources (threats to employment). But the local management board was also able to mobilise political resources: it succeeded in gaining support from the workers of the firm and from institutional bodies (mainly DRIR). The strategy of PEC moved slowly during these four years from confrontation - when the Company felt strong enough - to dialogue - when the local opposition weakened its position. Thus, one can say that the company as a whole acted as a director within this interaction network. Since this first phase, as during the whole decision making process, it always tried to promote a problem solving pattern of interaction, and to maintain the arena within technical and institutional limits. Therefore, when the conflict with the population occurred, it let the political and institutional actors solve it, after being ensured that these actors would follow its line and would support its project.

2. The second arena involved in the process is the political one, composed of elected bodies at local and national levels. Several interventions were organised by these elected representatives, in order to speed up the closure and the transfer of the St Maurice plant. The main promoter during this phase was the municipality of St Maurice which succeeded in mobilising the NGO 'against the PEC facility's pollution' in order to counterbalance the lack of intervention from the Prefectoral administration in favour of the pollutant plant. The main objective of the mayor concerned was at that time mainly electoral, in favour of the well-being of the St Maurice inhabitants. Faced by this political problem related to his electoral mandate (i.e. the concrete opposition of the population), the mayor of St Maurice was able to mobilise political resources (pressure on various institutions, through and outside the Communist party) in order to incite this intervention in the institutional arena.

3. The last arena involved is the institutional arena composed of many different units: the prefectoral administration (Préfet and sous-Préfet), the DRIR and the ministry of the Environment. The DDASS[4] (Provincial Agency for Health and Social Affairs) and the administrative Court are also involved in this process, although to a lesser extent. During this phase, these actors tried to mediate the conflict between the two main opponents (the company and the St Maurice inhabitants), especially with the emergence of a mediator in 1981 - the 'Sous-préfet of Vienne'- who is the representative of the Préfet of Isère at the local level, and whose action was a determining factor in finding a solution to the conflict. But some contradictions appear within this arena. On one hand, the DRIR disagreed with the transfer of the plant, whereas, on the other, the Ministry for the Environment - the D.P.P.: Pollution Prevention Department - tended to listen favourably to the arguments put forward from the social and political arenas. During this phase, the problem

[4] The DDASS, Provincial Agency for Health and Social Affairs (Direction Departmental de l'Action Sanitaire et Sociale), is a provincial agency of the Health ministry. In each French department, the DDASS checks the quality of the environment in relation with public health.

considered by the State administration was mainly a problem of waste treatment capacity: France, and especially the mostly industrialised Rhône Alps region needed waste treatment plants. However, at a local and regional level, then at the national level - as far as the opposition stemming from the population became more and more visible - the perception of the problem changed and turned into a pollution control problem. The intervention of the DDASS, which undertook an epidemiological study related to the health impact of air pollution emitted by the St Maurice plant, partly explains this change. Then, as a whole, the administration followed two main objectives: the protection of the environment (by means of waste treatment as well as by means of pollutant plant control) and the maintenance of public local order (in political terms). To that end, it succeeded in mobilising legal (formal authorisation and bans) as well as political resources (mediation), and developed a bargaining approach.

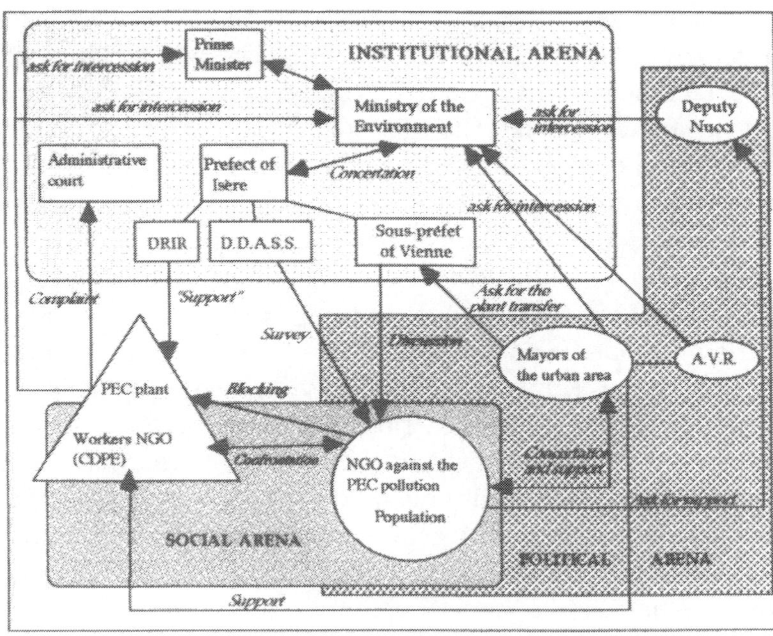

FIGURE 2.1: Local opposition to the plant at St Maurice (phase 1: 1976-81)

Phase 2 (1981-82): Solving the pollution problem at St Maurice

The previous conflicting period ended with the signature of the protocol. The technical and institutional arenas then prevail once again (cf. Chart 2). The main actors negotiate in a more formalised way, within an ad-hoc Committee, which was created in order to follow the setting up of the filter facility on the St Maurice Plant, and afterwards to follow the transfer of the plant to a new location.

The PEC management board and the State administration constitute the main actors of this phase, under the close supervision of the NGO 'against PEC pollution' and of the local elected bodies.

During this phase, the problem considered by the company is both an economic as well as a technical problem. More precisely, at that time, the relationship between the different levels of decision making within the company was not based on a strict hierarchical mode. The central and the local management boards acted as decentralised bodies, but on a complementary basis: the first one had an economic-strategic purpose (economic stake i.e. willingness to preserve and finally to expand the waste treatment activity of the firm) whereas the aim of the second one was more technical (willingness to promote its technical know-how, especially as far as the air filter is concern). In order to achieve this objective the company mobilised mainly economic (in terms of finance and employment) and cognitive (mainly technical know-how) resources.

The second important actor of this phase is the State administration. The administration mobilised economic and cognitive resources during this phase: on one hand the Sous-Préfet implemented its role of mediator, by formalising an agreement between all of the actors involved, and on the other hand the Ministry of the Environment, through the water agency's financial help,[5] brought financial means to the disposal of the company in order to install air filters on the St Maurice stacks. The problem considered by the administration is then the same as that in the previous phase: a problem of waste treatment capacity and a problem of pollution abatement. But the stake differed from one administrative unit to another. Whereas the central administration (the ministry for the Environment which was created in 1971), tried to gain more politico--administrative skill and power, the Prefect on the one hand and the DRIR

[5] The Water Agency is an independent agency in charge of collecting taxes on water polluters and which distributes subsidies to companies and local governments which invest in water protection devices. The six French water agencies are under the tutelage of the Ministry of the Environment.

on the other tried to get more political and technical recognition from their respective traditional partners.

During this phase, the local and departmental elected representatives tried to control the social movement, and to convince the NGO 'against PEC pollution' to accept the transfer of the plant rather than its closure. Thus, during this phase the stake for them was an economic one: the problem is first of all to safeguard or create jobs located in the area, but it is also to maintain satisfactory conditions for the regional chemical industries, which are the main employers of the region. And these industries needed waste treatment plants. In order to convince the NGO and to control the expansion of the local opposition, these elected representatives use mainly cognitive resources.

This phase constitutes the apogee of the NGO against PEC pollution action. It is formally recognised as a partner within the decision process: it is invited to sign the protocol. And it succeeds in stopping the air pollution emission, by imposing the air filter in the St Maurice plant, even if it compromised a bit by accepting the transfer the plant instead of its closure. Thus one can say that the objective of this actor was achieved.

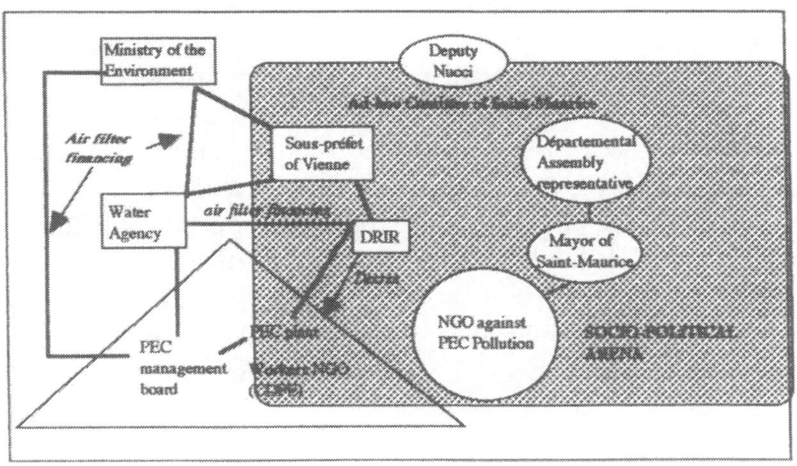

FIGURE 2.2: Solving the pollution problem at St Maurice (Phase 2: 1981-82)

Phase 3 (1982-84): The transfer of the St Maurice plant

During this phase the political and to a lesser extent the institutional arenas play the main roles. Three main issues are at stake during this phase.

The first one is about the search for a new location. We notice, however, that as far as this issue is concerned, the phase might likely have begun before the end of the previous phase. But due to the secrecy which qualified this informal process it is difficult to precisely define the start of this negotiation process. Nevertheless, this issue is dominated by the intervention of local elected bodies: mainly the mayor of St Maurice (who is at the same time the president of the SIVOM de Roussillon), the mayor of Salaise and the representative of the departmental council. During this phase the resources used by these actors are mainly political: the relationship between these elected persons went through the Communist party and through the SIVOM de Roussillon, which represents a formal structure of co-operation between local governments. The main objective pursued by these actors was to transfer the plant from St Maurice to a site located in the same area. As in the previous phase, their problem perception is linked with an economic stake. This is the reason why they adopted a problem-solving mode of interaction. But they also used cognitive resources in order to retain control over the participation and the information of the local population, by preparing the decision at a political level before opening the discussion.

The second issue is a technical one. It concerns the technical demand which is set up in collaboration between the DRIR (Regional Agency of the Ministry of the Environment and of Industry) and the PEC-TREDI management board, but with the support of local elected bodies involved in the ad-hoc Committee of Salaise (created in order to prepare and follow the building up of the new facility). The main results of this process are the drawing up of guarantees for the functioning and the control over the new plant. During this phase the administration used mainly legal resources through the implementation of the pollution control regulations, but also cognitive resources, through the technical skill of the DRIR engineer. Thus, the role of the administration can be considered as a gatekeeper mainly played by the DRIR, without which the decision would not succeed. Such a role requires a bargaining approach. The objective of the DRIR at that time was to prevent the pollution from the new plant by the implementation of the best available technology, the feasibility of which was meanwhile demonstrated at the St Maurice plant. Whereas, during the same period, the objective of the company was to ensure the management of the new plant (on technical and economic terms) and also to ensure that the size of the new plant would be bigger than the St Maurice one.

The last issue is about the financing of the transfer. It is dominated by an institutional arena composed mainly of the ministry of the Environment and the Préfet of Isère. Their intervention was a determining factor in finding subsidies and low rate credits for the new PEC-TREDI plant. However, the local elected bodies also play an important role inasmuch as they directly participated in the financing of the transfer. Then the resources mobilised by the administration and the local elected bodies on this issue are mainly cognitive and economic.

During this phase, the mode of interaction between actors is clearly oriented to problem solving. And the mayors finally appear to be the most important actors, playing a role of promoter, since they intervene on each issue of the process whether diplomatic or technical or financial. The social arena - NGO 'against the PEC facility's pollution' and inhabitants of Salaise - is not really involved during this phase, except when a one year's delay was requested for the transfer, and when a new conflicting situation occurred at St Maurice. The NGO against the PEC facility's pollution first of all refused to accept this one more year's delay, but in the face of the coalition of other actors and the risk of being considered as responsible for redundancies, it accepted.

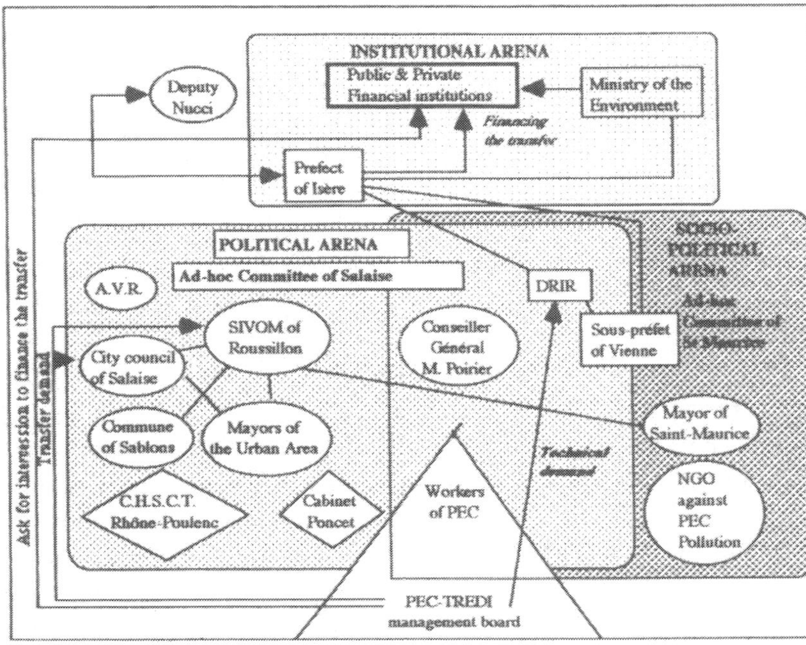

FIGURE 2.3: Solving the problem at Salaise (Phase 3: 1982-84)

Phase 4 (1985-91): The new TREDI-Salaise plant in operation and its extension

During this phase the technical arena is the only one in operation. The City council of Salaise remained informed of the technical developments of the project, but it did not play an active role.

As imposed by the national law and the decree on classified installations, the main actors which must conduct these kind of project are the DRIR on behalf of the Préfet, with the help of private technical agencies like ALPAVE (a private agency recognised by the administration) and the management board of the plant. Later on, the plant was extended (1988-91); then during this period the technical area was slightly opened to a political one as the national regulation required. During this phase, the management board of the plant took the initiative and the direction of the interaction process. But in order to perpetuate such a collaboration between technical and political actors, a new committee was created at a regional scale, in order to jointly manage technical information about the pollution stemming from industries settled in the urban area of Salaise.

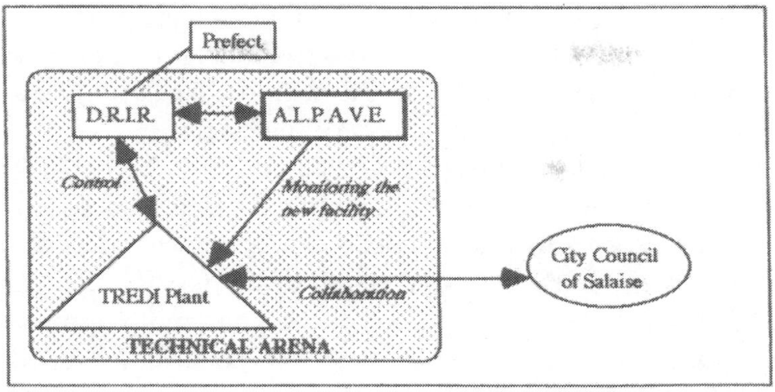

FIGURE 2.4: The new TREDI-Salaise plant in operation (Phase 4: 198591)

2.4 THE SUCCESS FACTORS OF THE FRENCH CASE

Five main types of success factors have been identified. Some refer to the socio-economic context and the socio-political context, others refer to the location and technical content of the project, and to the value added by the project; the last ones referred to the decision making process itself (see Chart 5).

(a) The socio-economic context
The socio-economic context of the case played an important role. As a matter of fact, the economic situation in terms of unemployment in the region and the municipalities concerned was able to counterbalance the nuisance problem. The safeguarding of the jobs, both for the waste incinerator itself as well as for regional chemical industries (which might be affected by the absence of a waste treatment facility within the region) was a strong argument for transferring the plant rather than closing it. As a matter of fact, the safeguarding of the jobs related to the waste facility (less than 20 jobs) was at least of symbolic importance.
(b) The socio-political context
Two factors can be presented as important for the socio-political context. The population of these villages is composed mainly of people working in the chemical industry, with a strong working class culture. This working class culture, which impregnates both the local population and the local governments, helped in the acceptance of the project. Thus the

adherence of the local mayors and elected representatives to the working class, and their previous involvement in local chemical factories - before retiring, most of them worked for the Rhone-Poulenc company which is the most important waste producer in the region - partly explained their ability to adopt a problem solving pattern of interaction.

More obvious is the membership of elected actors in the Communist Party. Such membership induced two combined types of behaviour: first of all, a very close linkage between elected persons and 'their' population, which explains the mobilisation of the population against the plant at St Maurice, and the acceptance of the plant by the population of Salaise. Second, these actors acted mostly in an 'inter-communal' mode, which counterbalanced the natural NIMBY syndrome which does exist at the local governments level.

(c) Factors related to the location and the technical contents of the project

As is already known, some success factors relied on the location of the site.

First of all the location of the waste treatment plant within an industrial region, seems to play an important role for two main reasons: the proximity of the waste producers, and the ability of the population to accept industrial plants (see above).

At the project level, the choice of a location within an industrial estate, specialised in chemical factories, peripheral to urban areas and with an already good level of road infrastructure helped the acceptance from local, concerned people and also from elected actors. As a matter of fact, the location chosen was proposed because it complied with these criteria.

Moreover, some factors related to the technical side of the project have probably played a role. The Salaise project appeared to integrate the best available clean technology of the period. And above all, the positive environmental impact of this technology had already been demonstrated at the previous St Maurice site. Such a demonstration was an important argument, convincing the mayor of Salaise to accept the settlement of the facility in the commune of Salaise territory, and was also an argument used by the Mayor of Salaise to convince the population of the low impact of the new facility. Moreover, such clean behaviour was important to the mayor of Salaise in order to allow another industrial plants to settle into the new industrial zone.

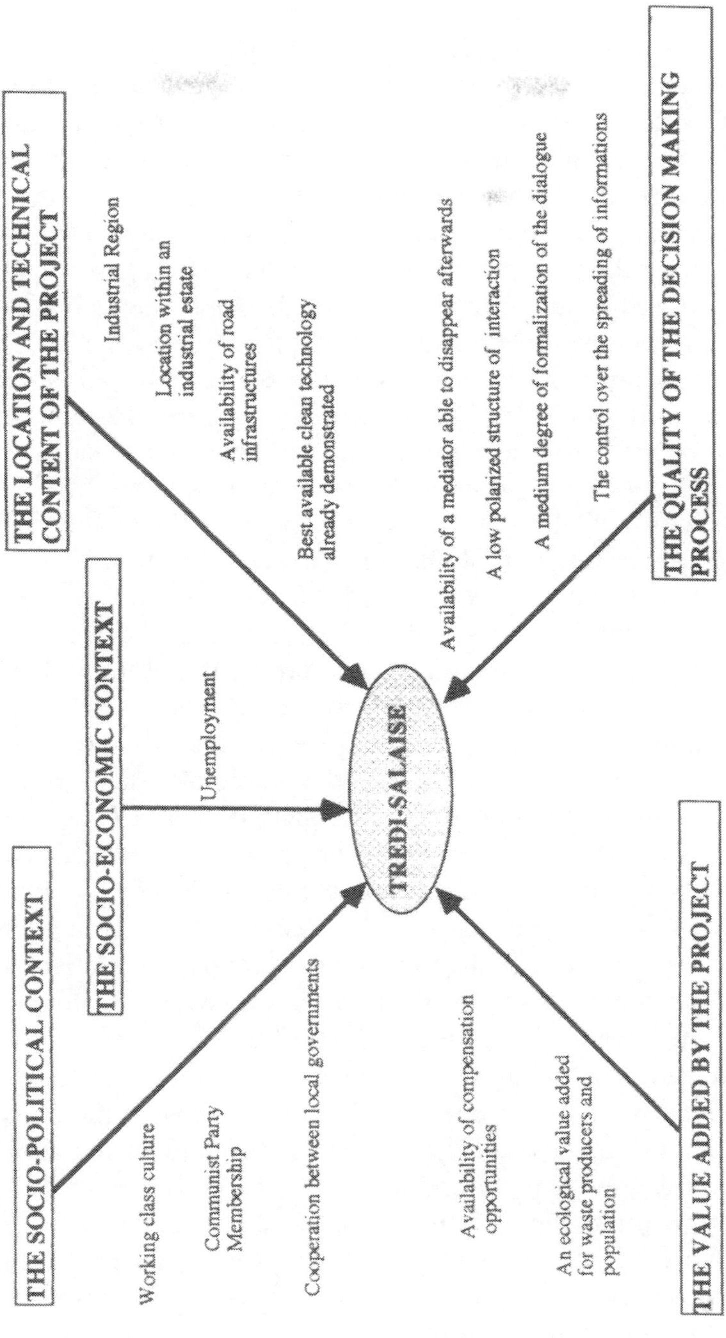

FIGURE 2.5 : The success factors of the French TREDI-salaise decision making process

(d) Factors related to the added value of the plant

The value added by the project is both an ecological and economic one. The TREDI-Salaise case points out that one important success factor is related to the ecological value added by the project. In the case of TREDI-Salaise plant, this ecological added value was valuable in two instances. First of all, when the St Maurice plant was set up to replace a nauseous landfill, and second, when the TREDI-Salaise plant replaced the damaging St Maurice plant. And the demonstration of this ecological added value before the transfer of the plant helped the success of the decision making process. Then the project represented an alternative solution to pollution that has been already emitted: the perception of the problem changed from a pollution control approach towards a pollution prevention approach. Rather than searching for new sites, the administration and the companies in charge of the setting up of waste treatment plants ought to look for old nauseous landfill or damaging waste treatment plants in order to replace them by cleaner ones. Second, the proximity of the waste producer induced a demand for waste treatment located in the region; such a proximity let the waste treatment side of the problem prevail, not only for the Environmental Administration, but also for local governments; all the more because these local governments have to face the problem of their domestic waste treatment, which are treated by the TREDI plant. Thus the pressure emanating from the demand side occurred both for industrial wastes and domestic ones. Such statements tend to confirm the positive impact of the availability of compensation opportunities: the environmental costs borne by local governments and the inhabitants were partly compensated by the benefit they received in terms of domestic waste treatment (cheaply treated by the new Salaise plant), as well as means of industrial waste treatment and their related safeguarded jobs. As a matter of fact, two types of jobs were in the balance: the jobs created or maintained in the waste treatment plant itself, and the industrial jobs of the region, which could be endangered if the industrial wastes produced by the chemical plants lay without any treatment, in a period when waste regulation became increasingly stringent. Moreover, one should notice that such a compensation opportunity was possible because there was no competition between the communes involved within the process about the local tax revenues. The city council of St Maurice has not lost the local tax imposed on the St Maurice plant, because the tax paid by TREDI-Salaise benefits the inter-communal body (the SIVOM de Roussillon).

(e) Factors related to the decision making process
Three factors must be pointed out.

1. First of all, the course of the decision making process itself. We identified several stages of the decision making process (see Section 2). Two main remarks can be pointed out from this chronology. First of all, whereas the decision making process began and ended within a technical arena, a socio-political arena prevailed in the middle of the decision making process, i.e. when the conflict occurred. The ability of the mediator to disappear after the conflict period and after being ensured that the protocol would be respected, contributed to the success of the story.

The second remark is related to the structure of the network: during the decision making process the structure of the network had changed from an axial to an interactive structure. Such a change can be attributed to the pressure coming from the opponents, who always tried to open the network. This confirms the positive impact of a low polarised structure of interaction during the decision making process. The interaction network in the case of Salaise reveals a balanced constellation of actors, where none of the involved actors were able, alone, to block the process or to end it. Therefore, the decision making process moved from a conflicting and a bargaining mode of interaction towards a problem solving pattern of interaction. Such a change, which explains and reflects the success of the story, can be linked with the low level of polarisation of the network.

2. The second point which can be stressed as regards the decision making process is related to the degree of formalisation of the dialogue between the actors. The dialogue between the different actors rapidly became formalised during the decision making process, the best manifestation of which was the signature of the 1981 protocol. This formalisation activity was the best means to obtain the confidence of the opponents and of the local governments in the behaviour of the administration and of the Company. But this formalisation occurred during the decision making process and was not pre-defined by any legal procedure. The protocol does not have any legal value as such; but it has a very important symbolic value.

The dialogue was also formalised during the stage related to the transfer of the plant, with the creation of the commission in charge of following the transfer (ad-hoc Committee of St Maurice). Likewise, a formal committee was created in order to control the running of the new plant (ad-hoc Committee of Salaise). Here again, such a Committee was not induced by the implementation of legal means. It was imposed by the mayor of Salaise as a formal guarantee of the good running of the facility. Here again the symbolic value of this formalisation is obvious.

3. The third factor is related to control over the spreading of information. The transparency of the decision making process was another important factor. Here again, the rules of the game related to the circulation of information among the different actors, was defined during the decision making process.

Such a definition occurred during the conflict stage related to the St Maurice plant, but also during the stage related to the transfer of the plant. In both cases, the control over the channel and the period of the spread of information was important. Not only the content and the form of the information were important, but also the moment of their availability.

2.5 CONCLUSIONS

To conclude, one can make some general statements about the French case, which might help following decision making processes related to waste facility siting. The first one is related to the economic development at stake during the whole decision making process. As a matter of fact, the problem to be solved at St Maurice and Salaise was not a problem of waste and nuisance, but a problem of economic and especially industrial development. For all the actors involved, including the population which formally acted against the pollution, the background problem of the process relies upon the maintenance of industrial activities within the region.

The second important factor is related to the involvement of the local elected bodies which have played a key role throughout the decision making process. Such a key factor was missing in the case of another

TREDI plant located in the North of France, and which has closed due to the absence of such support.

Thus, the socio-political and socio-economic factors, presented above as success factors, have been undoubtedly the most important factors to have contributed to the happy end of the story. Such 'cultural' conditions not only explain the success of the decision making process but also the perception of the problem itself. It led everybody to agree, implicitly or explicitly, to build a new plant. But such a structural variable stems mainly from the past. It dominated the opinion of political and economic actors during the 60s and the 70s.

This way of thinking is less predominant, nowadays except, maybe, within the Communist Party, which explains why most of the new waste treatment plants were built in communes governed by politicians belonging to this Party, and also in industrial regions.

For the future, such success factors can hardly be proposed as such, because such structural variables are less and less framing the economic development of France. Moreover, as the economic situation of the waste industries improves in France, the financial help which was importantly provided to TREDI by various public bodies will no longer be offered at least in such an important way. The quality of the decision making, the technical content of the project and above all the value added by the project might then be strengthened for new waste treatment plant decision making processes.

3 THE CONFLICTS ENGENDERED BY WASTE FACILITIES SITING: LESSONS TO BE LEARNED FROM A SPANISH CASE

SUSANA AGUILAR, JOAN SUBIRATS

3.1 INTRODUCTION: THE PROBLEM

The case under analysis (actually a dual case since it involves a success and also a failure) takes place in the region of Andalusia, more specifically in the province of Huelva. It is exemplified by the building up of a waste facility (micro-encapsulating plant) in Palos de la Frontera in 1992, after a linked and failed experience regarding the attempt to install a landfill in Gibraleón.

Huelva is one of the most heavily polluted provinces in Spain. The province has often undergone serious episodes of air pollution and, in 1980, it ranked third (after Murcia and Vizcaya) in sulphur dioxide (SO_2) emissions in Spain. On the whole, most pollutants come from industrial processes (mainly from the chemical sector).

All this explains why this province is one of those covered by the, still embryonic, network for the surveillance of industrial pollution (Figure 1).

B. Dente et al. (eds.), The Waste and the Backyard, 69–98.

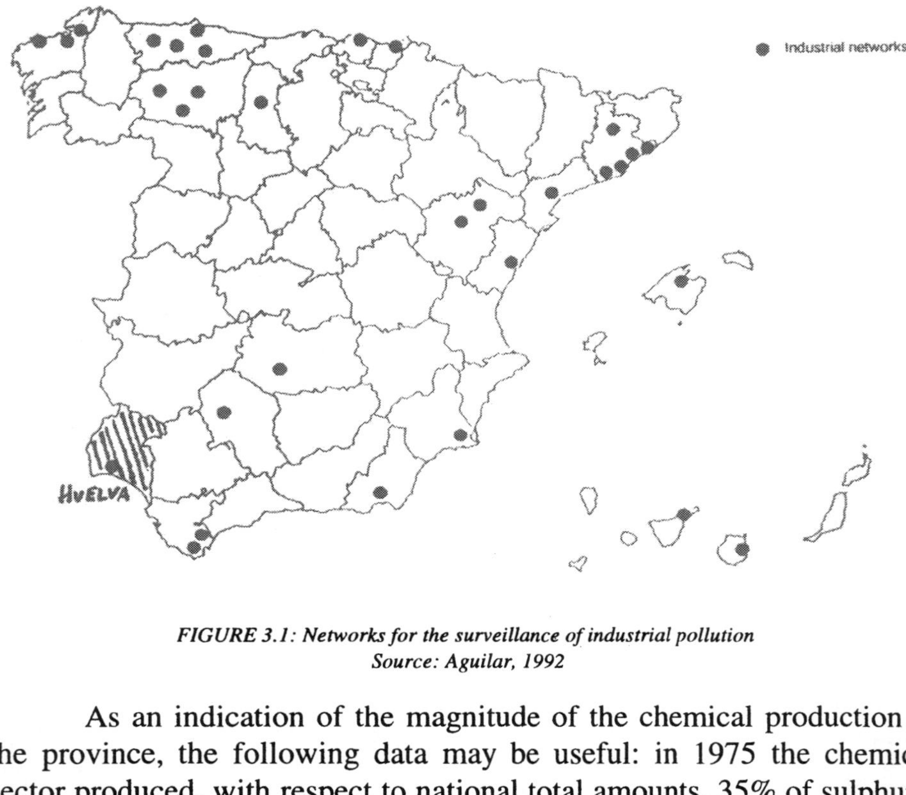

FIGURE 3.1: Networks for the surveillance of industrial pollution
Source: Aguilar, 1992

As an indication of the magnitude of the chemical production in the province, the following data may be useful: in 1975 the chemical sector produced, with respect to national total amounts, 35% of sulphuric acid, 96% of phosphoric acid, 100% of polyphosphates, 28% of ammonia, 52% of benzene, 100% of cyclohexane, and 100% of cellulose. Along with this production, the chemical activities led to the generation of, among other residues, 765,000 tons pyrite ashes per year, and 2,700,000 tons calcium phosphates per year. These quantities significantly contribute to the 1,700,000 tons industrial waste annually generated in Spain (of which only 400,000 tons are duly disposed of).

The first important attempt to face the environmental problem in Huelva started out in the mid 80s with the implementation of the Global Abatement Pollution Plan, designed by the socialist regional government.

(PSOE).[1] This plan embraced, among other things, the building of a publicly owned landfill for industrial waste. The first site chosen for the landfill was a town called Gibraleón which was ruled by a socialist mayor. This decision was taken after several technical reports coincided in the appropriateness of the siting. But, from the very outset, the population fiercely opposed this waste facility. The project unleashed impressive turmoil that led to tough police interventions, and to very serious political strife. Only when the actors participating to the policy process (mainly through a forum called the Clean-Up Table) became aware that social opposition was insurmountable was a decision made on a different site in Palos de la Frontera. This change was justified as a result of the discovery of a sophisticated waste facility (micro-encapsulating plant) which rendered the landfill obsolete. Although inter-party struggle then gradually faded away, some social resistance to the new plant arose initially in Palos. Despite this, the opposition lost its initial impetus with the passing of time and the waste facility was finally built in 1992. Since then it has functioned in a smooth way, although it continues to be underused to date.

---

[1] Since 1977, the date of the first general elections, the socialist party (PSOE) has been the most important political force in Andalusia (Annex A): In that year the PSOE received 35% of the votes compared to 29% at the national level. In 1982, when the first regional elections took place, the PSOE won 52% of the votes. Since then, the party has been in power and enjoyed an absolute majority in Andalusia, except for the 1994 elections in which votes dropped from 49.6% (62 seats) in 1990 to 38.7% (45 seats). The socialist predominance has also characterized politics at the central level until the last general elections held in 1993, in which the PSOE lost its absolute majority (159 seats out of a total of 350). The prospects for the forthcoming 1996 elections, to be held on both the national level and the Andalusian level, seem to further reinforce this decline since the PSOE is likely to fall from power.

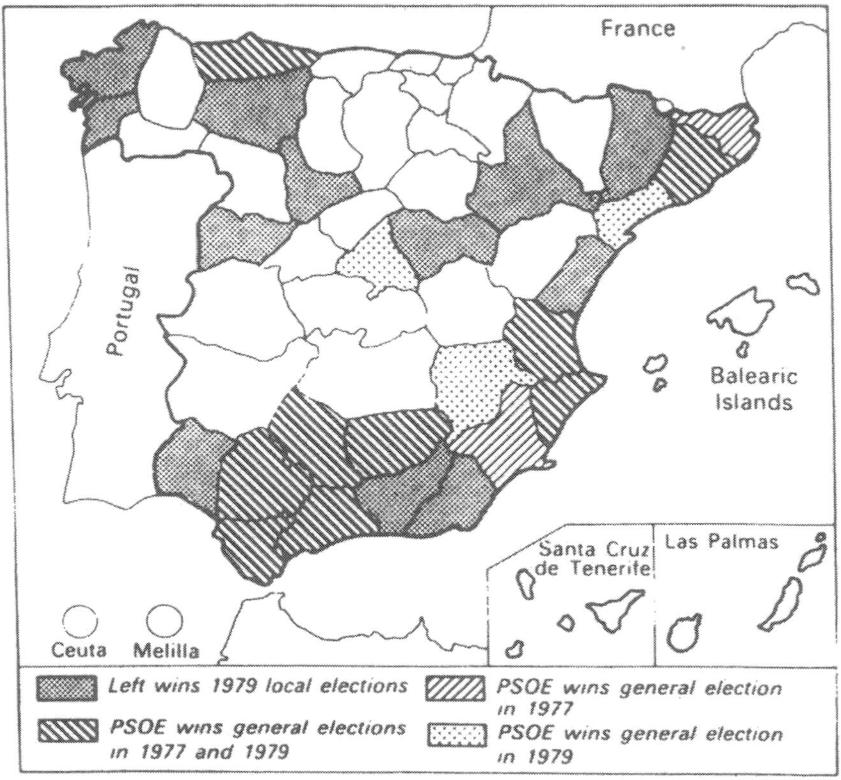

FIGURE 3.2: The socialists' electoral strongholds, 1977-1979

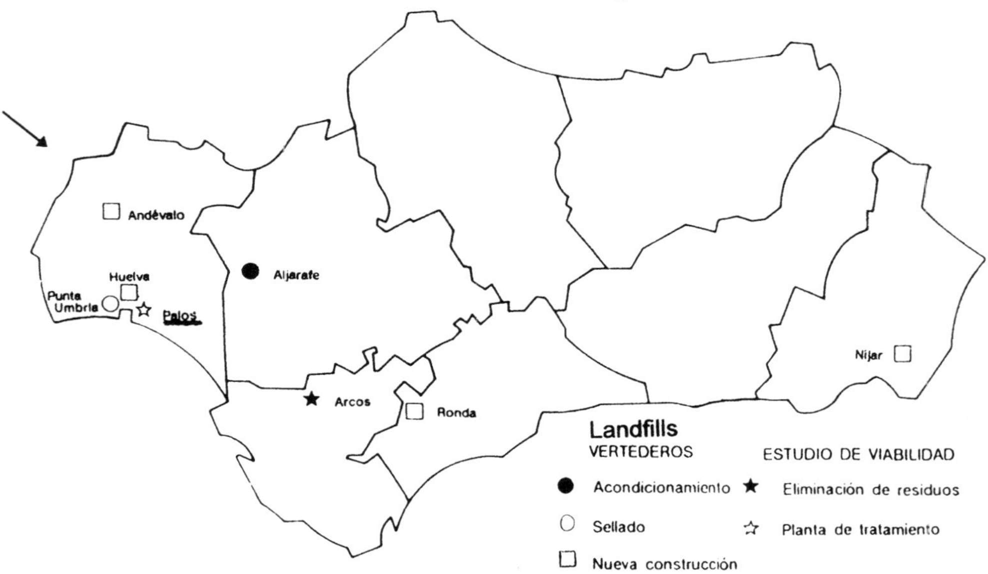

FIGURE 3.3: Location of the micro-encapsulating plant in Palos
Source: Junta de Andalucía, 1990

3.2 THE CHRONOLOGY

PHASE 0. *Background*
The province of Huelva, whose capital (also called Huelva) has 140,000 inhabitants, generates 80% of the industrial waste in Andalusia. In 1986 a Global Abatement Pollution Plan was designed, covering public measures (to be undertaken by the regional government -Junta) as well as measures which fall under the responsibility of private firms. The public measures included the building of a landfill which would allow the storage of the industrial residues generated in the course of the clean-up programs.

PHASE 1. *The Onset of a Conflict: Disputes around the Global Abatement Pollution Plan and First News about the Building of a Landfill (Feb. 1987-June 1988).*
(Feb-Nov 1987) The Junta announces that the firms which do not comply with the deadlines for the submission of clean-up projects will be sanctioned. A newspaper headline reads: Industrial pollution seriously deteriorates quality of life in Huelva. The ECOH organises a demonstration against pollution.

(Dec. 1987) The AMA buys a rural estate, 14 km away from the town of Gibraleón (G),[2] as future site for the landfill foreseen in the Global Abatement-Pollution Plan.

(June, 1988) News about the landfill leaks out to the public; first announced by the AMA, and later on defended by a PSOE elected local official. A person within the AMA points out that, in order to avoid protests, the definitive siting will not be publicised until the works begin; he also denies that the landfill will be constructed in G. A UGT member affirms that the AMA director once said that the landfill should be located within a distance of 15 km from the chemical concentration in Huelva, and this condition was not fulfilled by G. Likewise, a consulting firm verifies that there are four wells in the vicinity of the planned site for the landfill, and that waterproofing measures should be taken.

---

[2] Gibraleón is one of the 78 towns of the province of Huelva: In 1981 it had 9,044 inhabitants, mostly employed in agricultural work, and has an area of 329 km^2.

PHASE 2. *The Unleashing of a Conflict: Attempts at Building a Landfill and Social Opposition in Gibraleón (June 1988-Jan 1989).*
(June, 1988) The inhabitants of G. demand that the Town Council define its stance concerning the landfill, and demand the resignation of the mayor. The socialist mayor of G. has always maintained an ambiguous stance concerning the project: on the one hand, he has defended the goodness of the landfill as part of the Global Abatement Pollution Plan designed by the socialist regional government, but, on the other hand, he has tended to side with his constituency whenever opposition to the waste facility has arisen. The PP speaker in the City Council demands a campaign of information on the issue. Official statements claim that G. is the best place to locate the landfill, although no decision has been taken yet. It is leaked out that the AMA has already begun work in G., and the PP (within the Town Council) asks the PSOE to stop the work.

(Aug. 1988) An assembly of the CBL, created in G. to oppose the landfill, is held in the town. The mayor of G. remarks that the Town Council will not continue with the proceedings to be gain the landfill. The PSOE Executive Commission in Andalusia defends the plant in G. Protests and demonstrations against the landfill are organised in G. The mayor of G. expresses his solidarity with the neighbours.

(Sept. 1988) It is announced that a renowned lawyer will defend the case of G. against the landfill. Three PSOE councillors in G. resign due to citizens' opposition to the landfill. A sit-in is organised in the Town Council of G., and the civil guard (guardia civil) called by the mayor goes to the scene of the conflict. The AMA recalls that the project in G. is not a competence of the Town Council but exclusively of the Junta.

(Nov. 1988) The ECOH and the CBL submit four allegations against the landfill. The AMA president declares that it is impossible to reconsider the siting of the landfill. The advisor to the Presidency on environmental issues ratifies the decision about the site of the plant, and publicly thanks the mayor of G. for the service he is paying to the province of Huelva. The Provincial Planning Commission of Huelva decides to authorise the construction of the landfill. The mayor of G. underlines that he will not grant any municipal license for the landfill. A PP member questions the change of attitude of the mayor. Demonstrations against the landfill, and road blockades are organised; violence is used upon the mayor and PSOE councillors, and some people are arrested. The FOE offers to mediate between PSOE and PP. It is leaked out by the PP that a PSOE member tried to bribe the party into the landfill issue behind the scenes: in exchange for this, the town would receive a number of compensations and the PP would become member of a monitoring commission related to the landfill. The E.I.A. concerning the landfill is published. A PSOE member explains before the Regional Parliament the reasons that led to choose G.: it was the final recommendation of exhaustive technical reports. It is also claimed that, in case the Town Council dismisses the opening license, the Junta will set in motion all the available mechanisms to put the project forward. The so-called 'Sad Night' occurs: 200 inhabitants of G. are injured after confronting the civil guard, and 13 people are imprisoned. The Governor Advisor grants the opening license for the landfill.

(Dec. 1988) The Commission of the Municipal Government of the Town Council of G. dismisses the opening license requested by the AMA. A newspaper affirms that, at a plenary meeting in G. (6th Feb. 1986), the PSOE agreed on a project concerning the waste facility. The mayor of G. announces that a decision will be made soon about the resignation of the Council team; this resignation will be followed by the appointment of a Municipal Management Agency. A PSOE deputy in Huelva assures that the plant in G. will be built soon. Since the Town Council has refused to grant the license, the Planning Commission will have to pass the project. In the last case, the Head Council of the Junta will ratify the project. A new crisis is unleashed in the Town Council of G. because the mayor announces he does not intend to resign.

(Jan. 1989) Creation of EGMASA. The AIQB calls for a compromise in G.

PHASE 3. *First Signs of Negotiation (the Setting of the Clean-Up Table) and the Shifting of a Conflict: The Substitution of Gibraléon by Palos de la Frontera as a New Site for a Waste Facility (Jan. 1989-July 1989).*
(Jan. 1989) A PSOE member declares that his party is ready to reach a compromise about G. without prior conditions. An AIQB manager states that, under the condition that a prior study is made, industry would not oppose the building of a landfill in Palos.[3] The FOE invites all the political forces and unions of Huelva to participate, along with the AMA, in the negotiation about the site of the landfill. After this invitation, the possibility of creating a negotiating table is discussed; the PSOE opposes the presence there of the CBL. It is also accepted that a consulting firm will produce a report on the most suitable place for a landfill. The PP quits the Clean-up Table when it does not get the AMA to accept a document that guarantees the termination of the proceedings concerning the project in G. The progress of work for the landfill is paralysed by the AMA, but the administration does not give up the idea of building the plant in that town if a committee of experts considers G. to be the best place. Two actions against the project presented by the lawyer defending G. are rejected by the Territorial Court.

(Feb. 1989) A PP member claims that the Clean-up Table, set up by the FOE, has no legitimacy. The Table entrusts the consulting firm Epypsa with the research on the landfill, but IU express its uneasiness because this firm has several times been contracted by the AMA.

(April, 1989) The Table decides to leave aside the project entrusted to Epypsa, and accepts the proposal of a PSOE member about the discovery of new pollution abatement technologies that could function as alternative to the landfill in G. The Junta considers the possibility of abandoning the setting up of the landfill in that town. Some AMA experts have visited several countries to analyse different waste treatment techniques. According to the AMA the best technology seems to be the encapsulating or coating plant, because it uses a sophisticated method for waste recycling and has already proved successful in some Western countries. The final decision on the landfill is therefore put off, and the Table decides to keep the proceedings concerning it paralysed.

---

[3] Fifty per cent of the firms belonging to the chemical sector of Huelva are located in Palos de la Frontera. This town had 5,798 inhabitants in 1981, who mostly work in the industries of the area and also in the agricultural sector, and its size is 48.7 km^2.

(May, 1989) The AMA assures that the new technique is about to pass the last tests: residues become totally harmless by means of an encapsulating technique developed by a multinational firm from the USA (demonstrated technology). A commission of the Table will travel to Italy to visit some waste facilities of this kind. The AMA president announces that the plant would probably be located in the industrial estate of Nuevo Puerto in Palos de la Frontera. The advisor of the Presidency announces, before the Regional Parliament, that the landfill at G. is no longer necessary. The AMA finally discards G. as the site for the landfill.

(July, 1989) The Table accepts Palos as the site for building an encapsulating plant. A newspaper headline reads: 'Landfill foreseen in G. will be finally located in Palos'.

PHASE 4. *Initial Resistance to the Building of a Micro-encapsulating Plant in Palos de la Frontera (July 1989-Feb. 1990).*

(July, 1989) A newspaper affirms that the resistance of Palos against the plant is due to previous false news, and it explains that no landfill is going to be built because the AMA and the Table have found another technology which renders it unnecessary. The Table also affirms that the initial opposition of Palos is due to the lack of information. The FOE president declares that he has tried to contact the mayor of Palos to give her information about the plant.

(Dec. 1989) The Town Council of Palos opposes the location of a waste facility in the industrial estate of Nuevo Puerto. The FOE secretary expresses his ignorance about whether the AMA had had contacts with the council of Palos; he also concurred with other participants in the Table in criticising the attitude of the locals. The Table gathers to set a date for informing the Council of Palos about the project concerning the plant. At a Table meeting the possibility was mentioned of not getting FEDER funds if the application was not submitted in time.

(Jan. 1990) The inhabitants of Palos mobilise against the plant. A farmers' commission asks for the requalification of the estate where the plant should be located. A UGT member in the Table considers that the conflict is irrational since the plant is harmless, and the factories already functioning in Palos are potentially more dangerous than the waste facility. The environmental advisor to the Town Council of Palos writes a report on the plant, in which it is stated that the waste facility will neither be harmful nor cause any damage to the town, but he laments that the method of deep injections (cleaner but more expensive and totally supported by the EPA in USA) is not foreseen in the project. The Table agrees, as requested by the Council of Palos, to concede a four-month prorogation in the proceedings leading to the opening of a plant. The agreement also embraces inviting the Council to visit the plant in Italy; to provide all the necessary information in case the visit should not be enough to clear doubts; and to request the administration to make the procedures more flexible if further information were needed. The Table gathers to analyse the situation created after the Council of Palos decided not to grant a license for the work. A newspaper criticises the lack of co-operation shown by Palos. The mayor of Palos, the municipal corporation, and the farmers of the town do not accept the invitation to visit a plant in Italy (Modena). The AMA declares that, in the last instance, the Table will make the decision. The Council of Palos gathered with the Table and a representation of the City Council of Huelva, commits to study the project of the plant after receiving information by the AMA. The Council of Palos also agrees that the plant has nothing to do with the projected landfill in G. The administration wants to speed up the bureaucratic proceedings framed in article 180 of the Law on the Soil; according to it, the step to be followed - after the refusal of Palos to the plant - is to submit the project to the Department of Public Works and Transport of the Junta, which will proceed to send it to the Chief Council of the Junta.

(Feb. 1990) Palos de la Frontera will have a waste facility in spite of people's opposition. According to the AMA, the Council must by law grant all the licenses because the site belongs to an industrial area. The Town Council of Palos does not grant the license of activities to the plant, as it did not grant the license for work either. Given the interest in the project, it is likely that the Chief Council of the Junta will have to finally approve the project. The AMA recently submitted a letter to the Council stating that it had not asked for the granting of the license of work but for a study on the planning feasibility of the project; this is explained by means of art. 180.2 of the Regulation on Hazardous Activities which states that, if the request for the construction of a plant is made by the administration as a matter of urgency, and the plant is considered of public utility, there is no need to ask for a license for work. The Table affirms that the Planning Commission has approved the construction of the plant in Palos. A responsible member of the Department of Public Works says that the mayor of Palos explained, before the Planning Commission, that she was supporting people's demands for not constructing the waste facility, although she acknowledged the adequacy of the plant. The members of the Table express their satisfaction because the procedures followed have totally respected the legality and protected the economic and social interests of the town of Palos.

PHASE 5. *Final Solution of the Conflict: Vanishing Resistance and the Building of a Micro-encapsulating Plant in Palos de la Frontera (Feb 1990-1992).*

(Feb. 1990) The opposition to the plant is gradually becoming weaker. This waste facility is backed by political parties, unions, entrepreneurs, and some environmentalists. The mayor of Palos, who has been accused of unwillingness to get involved into the issue, simply seems to be fearful that the conflict of G. will be reproduced in her town. But, contrary to the previous case, the inhabitants of Palos are in a different position: many work in the nearby industries and, consequently, are more used to coping with pollution and residues. Therefore, their opposition is mainly based on their unwillingness to receive waste from other areas.

(June 1990) The Town Council of Palos appeals against the decision on the waste facility: Two appeals (3534/90 B y 3536/90 B) are presented before the Regional High Court.

(Oct. 1991) The Table decides to visit the work under way in Palos.

(July 1992) The plant of Palos is finally built but it is underused (at approximately 50% of its capacity). Its construction entailed an investment of 600 million pesetas, paid by the AMA and the FEDER funds.

(Dec. 1992) The members of the Table express their willingness to appear before the Parliamentary Committee on Waste Management to offer the experience accumulated by the negotiating forum. The Table also proposes to analyse the incorporation of the PP into the forum.

In March 1993, after a year of performance, the AMA publicises the data on the output of the plant in Palos. The plant is ready to dispose of 60,000 tons waste per year, out of the 200,000 tons produced in Andalusia. The AMA president affirms that the region still needs an industrial landfill.

3.3 ACTORS' ANALYSIS, NETWORKS AND PATTERNS OF INTERACTION

(1) The Onset of a Conflict: Disputes around the Global Abatement-Pollution Plan and First News about the Building of a Landfill.

During this first phase the problem is defined in a technical way: a waste facility is needed to dispose of the residues generated in the course of the clean-up measures undertaken by the firms of the chemical pole in Huelva. This technical definition is accompanied by a technical solution: the waste facility needed is a landfill to be built in an area which has proved to meet all the necessary technical (not social) requirements (Gibraleón). Such an 'aseptic' view was systematically upheld by the Andalusian Environmental Agency (AMA).

The AMA was the main promoter of the landfill in the policy process and showed, consequently, a clear content objective: the agency committed itself to solving the problem of the waste produced by the industries during the implementation of the Global Abatement Pollution Plan. In order to accomplish this aim, the AMA sided with two 'natural' allies at first: the socialist mayor of G., who followed the dictates of the regional government in different ways (more openly and enthusiastically at the beginning and more reluctantly later on, once the conflict had been unleashed and he feared to lose his constituency support), and the chemical firms. The AMA was, therefore, the main actor in this early stage and its presence overshadowed that of other participants in the policy process who were unaware of the repercussions associated with the leaking to the public of news about the construction of a landfill - in fact, most participants were taken by surprise when the conflict suddenly broke out. Only the PSOE, apart from the AMA, played a visible role at the outset: its role was that of a director who had a content objective and, for this reason, its performance could be considered an extension of that of the promoter (the AMA).

(2) The Unleashing of a Conflict: Attempts at Building a Landfill and Social Opposition in Gibraleón.

In the second phase, and once a tough resistance to the landfill began to take shape, the interaction among the different actors became more complex (in the sense that every actor interacts with the rest, although in different ways and with varying frequency) and blurred (the number of interactions grows and these lack any type of regular pattern or degree of institutionalisation).

The issue, as in the first stage, keeps being defined in technical terms by the AMA and the PSOE. Following a technocratic approach based on the 'one best solution', the landfill is defended as being both the most suitable and rational method to dispose of a shared problem and the best way to promote the 'general interest'. It is not understood by these regional actors that the problem definition of the locals (alongside other groups who support their protest) is different because they do not live from the industrial sector on the whole. Therefore, the main problem encountered by the promoters and directors of the project was that, firstly, the AMA simply relied on its important cognitive resources (expressed in scientific knowledge and expertise assets) whereas, secondly, the PSOE overused its legal resources (in terms of knowledge about formal procedures) and political power in order to impose a decision which finally had to be abandoned. Even when it was evident that this plan could not succeed, both actors continued stubbornly to defend its adequacy for some time.

As regards cognitive resources, the AMA counted heavily on its own experts.[4] During the policy process they made use of their scientific knowledge, and were also sensitive to economic arguments (the costs of alternative techniques). The experts working within the AMA usually acted as filters, that is, as carriers of other actors' requests: the request of the AIQB for a quick solution, the request of the environmental administration itself as promoter of the waste facility, etc. In this sense, their pattern of interaction is basically based on problem-solving and the policy process for them is defined as the framework in which the best means for a certain goal has to be found (technocratic approach). Therefore, they tend to see the process as a positive sum game in which all actors get benefits from a (good) technical decision. This is also the reason why they do not usually feel comfortable in political arenas, and tend to underestimate the importance of ('non-objective and biased') institutional and political factors in order to reach solutions.

This 'technocratic blindness' on the part of the AMA and its experts, coupled with a certain feeling of prepotency on the part of the Junta because the region had traditionally been a PSOE stronghold (the party had been in power since 1982 and had retained the absolute majority in Andalusia until 1994), accounts for the attempt to impose the

[4] Experts can nonetheless work as independent advisors, as the case of Benito de la Morena exemplifies.

'best solution' and for the concomitant radicalisation of the stances of the different contending parties. In this radicalisation the role played by the regional PSOE is crucial. With the aim of helping to implement the solution of the landfill, the party extensively and intensively utilised its excellent political resources (in terms of links with most of the political and administrative institutions of the region), but miscalculated its strength. The attempt to impose the landfill, regardless of other parties' stance and citizens' protests, clearly ended up in a confrontation approach: The PSOE converted the project into an 'issue of pride'. For some time, this actor also regarded the problem as a zero-sum game since the concentration of resources at its disposal was so disproportionate from the outset that only winners and losers could be expected out of this process.

In this second stage another significant actor is the PP. The Popular Party played basically, within the realm of the political parties, the role of opponent. This party underlined that the landfill was not the best solution to waste disposal (objective content), although its main reason for opposing the project was, allegedly, the imposition approach followed by the Junta. Linked to these criticisms, there was a clear process objective (partisan reason): the PP tried to gain political support in a region whose reins had traditionally been held by the PSOE. The benefits were, however, limited, and most actors participating in the policy process criticised the exit strategy (Hirschman, 1970) followed by the PP: it was the only important party in the region which voluntarily remained outside the Clean-Up Table. For this reason, and despite the fact that the PP emphasised a problem-solving approach, it was not able (or willing) to link this approach with a bargaining pattern of interaction. The emphasis upon the biased character of the Clean-Up Table contributed to alienating the party from their 'natural' allies: the FOE and the AIQB. Likewise, its voluntary exclusion entailed that the PP confined itself to having some leverage only during the stages prior to the functioning of the Table whereas, afterwards, it had to limit itself to expressing its outrage as an outsider and to making criticisms which fell on deaf ears. Contrary to the PP, IU (or, more precisely, its representative José Quintero[5]) had a relevant and constructive stance from the early stages of the policy process. The importance of this figure, however, was

[5] That is, in this case it is better to talk about the individual and not the party as collective actor.

enhanced in the third stage, once the Clean-Up Table began to be summoned regularly.

Initially, the FOE had an important content objective (that is, the federation was interested in the implementation of a waste facility) because it represented the AIQB. Alongside this objective, the FOE pursued an strategy of mediation in the process in order to acquire prestige and consolidate its presence in other policy areas (process objective). Its role of mediator was clear in a dual sense: in its initiatives leading to the setting up of the Clean-Up Table, and in its chairing of the forum. The Association of Chemical Industries of Huelva (AIQB) did not exhibit a very visible role during the policy process because, as an important FOE member, the Federation acted as the carrier of its requests. Besides, this actor had no need to actively put forward its interests since it often found someone who defended them directly (the FOE), or indirectly (the PSOE, when opposing the relocation of the chemical sector, and IU, when defending the building of a waste facility to preserve industrial jobs). Basically, the Association had a problem-solving pattern of interaction, because a waste facility was needed to treat the residues generated in the course of the clean-up measures applied by the industries. However, if this solution could not be implemented, that would not be their fault (it was a compromise of the administration) and industries could always blame the Junta for not having accomplished this goal; besides, that failure would allow the firms not to internalise waste disposal costs.

Environmentalists and citizens (within the action committee) are better described as a social arena in which different actors and groups pursue shared objectives of a content character: the abandonment of the project to build a landfill. This objective led them to consider the situation as a zero-sum game and, therefore, they favoured a confrontational approach to the problem. They had a major role during the first stages of the conflict in G. where multiple demonstrations and road blockades were organised, but lost some visibility in the course of the policy process. Since they were never invited to participate in the Clean-Up Table, they were constrained to act out of the bargaining process. These two groups were the real social opponents to the project and, to reach their goal, they did not hesitate to side with, in principle, 'unnatural' allies like right-wing political parties (PP), and the clergy (the priest of G). They mainly used the strategy of fear, stressed the uncertainties which surrounded the project, and took advantage of some legal mechanisms (lodging complaints before the courts).

3) First Signs of Negotiation (the Setting of the Clean-Up Table) and the Shifting of a Conflict: The Substitution of Gibraléon by Palos de la Frontera as a New Site for a Waste Facility.

In the course of the process (more precisely in the third stage), the definition of the problem underwent a change: the problem ceased then to be the implementation of a good technical solution by all means available and became the need to persuade people about the goodness of that solution. In order to achieve this aim, the different actors came to accept - although with different degrees of conviction - that political measures should also be promoted (like the experiment of the Clean-Up Table).

The Clean-Up Table has proved to be a political and institutional arena of utmost importance in the resolution of the conflict. Although contending positions were present at the Table, the actors participating in that forum accepted the rules of the game connected with bargaining and the reaching of consensual solutions. This arena was not, however, all-encompassing: the PP left the Table when it was constituted, whereas the environmentalists and the citizens' movement were not permitted to join it. In the latter case, the consensual mechanism set by the Table whereby participation was only offered to a limited number of actors who were, then, entitled to make decisions, marks an important feature of a more general Spanish policy style.[6] Although the resolutions agreed by the Table had no binding character but a symbolic force, all actors participating to the process felt themselves tied by them.

Once 'objective' technical solutions became more or less useless, the AMA lost some of its internal unity and cohesion as an actor. From this moment onwards, the agency is better described as an arena in which different and conflicting views about the measures needed to cope with the waste problem, as well as diverse interpretations and reasons for the failure in Gibraleón, coexist. For some time, however, the landfill project continued to be supported by commissioned technical reports by the regional administration. Only when it was clear that the proposed solution was impracticable, and that the PSOE was eroding its social support, did the party began to accept the need for bargaining. It then became a commonplace to talk about the need for an ecological pact -

[6] In Spain, if we examine the intermediation mechanisms in sectoral realms, a participatory structure in the decision-making stage which is confined to few actors is likely to prevail (Subirats, 1992).

yet, the conception of this pact was limited since bargaining was confined to the members of the Table (political forces, administration, unions, and entrepreneurs), and the party firmly opposed discussion with environmentalists and citizens. However within the PSOE as collective actor, and following the same path that has been described for the AMA, different positions began to take vivid shape: for instance, the party at the local level played a dubious role, openly supporting the dictates of the regional party at first and, finally, ending up as opponent to the project.

IU, embodied by Quintero, could easily have played the role of gatekeeper, blocking the process with his political and interactive resources (strong links with the unions). Instead of that, this figure followed an approach which mixed both the roles of mediator and director: mediator because he somehow acted as the representative of the productive interests (unions, workers, and even companies) and was always willing to find solutions within the Clean-Up Table, and director because of his interest in preserving industrial jobs and facilitating the functioning of the companies. There is a clear linkage between personal placement on the productive side (leftist ideologies of a pre-materialist type) and eagerness to resolve (or to search for ways out of) waste issues connected with firms' economic functioning and material welfare in industrial areas. All this explains why this figure combines the bargaining and the problem-solving approaches. He also managed, along with the unions, to broaden the stake so that the issues about the possible relocation of the chemical sector and the future of the industries of the province were linked to the landfill problem and discussed at the Clean-Up Table. Besides, this person was extremely receptive to technical and legal arguments. By contrast, Quintero's allies, the unions, did not have much importance in the policy process because they basically followed the strategy of the IU representative. It could be said that unions acted as filters since they were somehow carrying the requests of actors who did not participate in the Clean-Up Table: the workers of the industries in Huelva. For this reason, they mainly had a problem-solving approach (the building of a waste facility would help firms to comply with anti-pollution plans, and could ease their normal functioning) which was promoted through bargaining strategies.

The FOE clearly attempted to take advantage of its mediation in the conflict of G. In this sense, the Federation's defence of a bargaining pattern of interaction was not only linked to a problem-solving approach, which could ease the waste disposal issue, but also to a process objective.

The emphasis upon bargaining, along with the fact that the FOE was for some time the only actor which managed to keep in touch with the other interested parties, led to its chairing the Clean-Up Table. This strategy, based on reaching agreements and compromises, is connected with the criticism expressed by some FOE members about both the politicisation of the landfill project (the PP was somehow blamed for having utilised the landfill issue for electoral reasons), and the strategy of imposition followed by the Junta. The mediation among the contending parties was possible due to the political and interactive resources of the FOE: the Federation was one of the few actors which was legitimated to look for a consensual solution, and it had contacts with all the contending groups. For this reason, it managed to put together all parties with parliamentary representation, the regional administration (basically the AMA), the involved industries, and the unions.

The AIQB, aware of the difficulties of dodging the increasingly tighter environmental controls exerted by the AMA, in this third phase used a strategy based on time pressure to urge a quick solution in the Clean-Up Table. In this respect it is important to bear in mind that the AIQB enjoyed important political and economic resources (in 1989 the industries of the chemical sector in Huelva provided 6,400 direct jobs), and mainly favoured technical arguments in the discussion about waste facilities siting.

(4) Initial Resistance to the Building of a Micro-encapsulating Plant in Palos de la Frontera.

Following the debates organised by the Clean-Up Table, the AMA committed itself to discovering a more suitable waste-disposal technique than the landfill, as well as a new location for it. In this search process, a micro-encapsulating plant was discovered in Italy by the agency which then organised trips to the country to convince the participants in the Clean-Up Table about the excellence of the waste facility. Therefore, the AMA continued using technical arguments but came to recognise, finally, that there need not be one best solution, but various solutions which should be discussed (and agreed on) by the different parties. In this sense, the AMA as well as the PSOE gave way to a bargaining strategy, without explicitly recognising that the previous project concerning the building of a landfill had been a 'less good' technical solution and a political failure to be blamed on them.

The discovery of this micro-encapsulating plant was enthusiastically welcomed by all the participants in the Table: the IU representative and the unions, sensitive to technical arguments and concerned about the negative economic repercussions (in terms of lay-offs) associated with the non-fulfilment of the Global Abatement Pollution Plan, welcomed the new project with open arms; the FOE and the AIQB considered the new technique a step forward with respect to the previous project and quickly appreciated the benefits linked to it (in terms of image improvement); and the AMA and the PSOE began to glimpse the end of a protracted conflict which had had a harmful impact on their public image. However, when the site for the new waste facility (Palos) was publicised, and some suspicion arose on the part of the PP mayor of the town (as well as the council), the PSOE, and to a lesser extent the AMA, seemed to be ready to apply the same imposition strategy which had rendered such bad results in the conflict of G. The arguments used to defend the location of the plant were again exclusively technical, and the legal and political resources in the hands of the party were quickly mobilised to get the new project accepted. Although the 'technicalities' (in terms of type of ownership of the land on which the waste facility was meant to be built, for instance) surrounding the siting in Palos were easier to sort out or to avoid, the initial response by the locals strongly resembled the still fresh riots in G.

The resistance in Palos had initially to do with a campaign organised by ECOH and CBL which tried to change (and broaden) the stake by linking the previous landfill issue to the current plant in Palos. But this attempt was finally unsuccessful because the participants in the Table hastened to utilise this forum as an institutional (and also technical) framework out of which a compromise should be reached. Alongside this strategy, a problem-solving strategy was clearly pursued by all actors to the process.

(5) Final Solution of the Conflict: Vanishing Resistance and the Building of a Micro-encapsulating Plant in Palos de la Frontera.

The proximity of factories as well as the industrial base in Palos is crucial in explaining the final acceptance of the micro-encapsulating plant. Regardless of these situative variables, it is also important to take into account that the Clean-Up Table (once the initial moves on the part of the AMA and the PSOE proved to be inadequate in the fourth phase) held the reins tightly in the process. This forum had already established a good reputation, and presented itself as an institutional arena in which solutions which were both technically good and socially acceptable were searched for. The interaction in this last stage then became mainly bilateral, between the FOE (and more precisely, its President), and the town council of Palos (its mayor). Personal connections between the mayor of Palos and industrialists within the AIQB and businessmen within the FOE also contributed to alleviating the previous tension unleashed by the regional agencies.

ACTORS	ROLES	RESOURCES	STRATEGIES	RESULT
AMA	Promoter	Legal	First, imposition (conflict); later on, bargaining	Loss of credibility
PSOE	Director (extension of promoter)	Political (and also interactive knowledge)	First, imposition (conflict); later on, bargaining	Defeat at the municipal elections in Gibraleón. Loss of political influence but circumscribed to a small area
PP	Opponent and, later on, gatekeeper	Political (insofar as gatekeeper)	First, confrontation; later on, problem-solving	Success at the municipal elections in Gibraleón. Limited political gains
IU	Director and mediator	Political (and interactive)	Bargaining and problem-solving	Prestige, enhanced role
Unions	Filter (somehow)	No clear resource (may be interactive)	Bargaining and problem-solving	No clear result
FOE	Mediator and director	Cognitive (in the sense of interactive knowledge) and political	Bargaining and problem-solving	Prestige, enhanced role
AIQB	Director (basically passive)	Scientific knowledge, economic, strategy of time	Problem-solving	Image improvement (situative variable: economic crisis)
ECOH	Opponent, filter	(Some) Cognitive and legal	Conflict and (attempt at) bargaining	No clear gains: may be an increase in interactive knowledge
CBL	Opponent, gatekeeper-filter	Mobilisation and legal	Conflict	Exhibition of strength

TABLE 3.1: Roles, resources, strategies and results of the main actors in the process

	PHASE 1 (Feb.87-June 88)	PHASE 2 (June 88-Jan. 89)	PHASE 3 (Jan. 89-July 89)	PHASE 4 (July 89-Feb. 90)	PHASE 5 (Feb. 90-1992)
PROBLEM DEFINITION	The building of a landfill in Gibraleón as the last step in the implementation of the global abatement-pollution plan	The imposition of the ('best') technical solution: The landfill	The need for a compromise to avoid the resistance against the landfill	The discovery of a better technical solution and the need to avoid past mistakes in the new site	Get the acceptance for the micro-encapsulating plant in Palos as the inevitable last step in the production Process
LIST OF ACTORS	AMA, mayor of G, firms	AMA, town council of G, parties (basically, PSOE and PP), FOE, ECOH and CBL	Clean-Up Table (FOE, AIQ, PSOE, IU, AMA, unions)	Clean-Up Table, town council and mayor of Palos	Clean-Up Table, town council of Palos
NETWORK STRUC-TURE	Star structure: (AMA as central actor)	Total and blurred interaction	Axis structure: Square (AMA, PSOE, IU, FOE)	Axis structure: Triangle (AMA, FOE, mayor of P)	Axis structure(bilateral): Clean-Up Table, town council of Palos
POLICY ARENAS	Technical arena	Political arena	Institutional and political arena	Institutional and technical arena	Technical and institutional arena
PREVAI-LING MODES OF INTER-ACTION	Problem-solving	Conflict	Bargaining	(Slight) confrontation and problem-Solving	Problem-Solving

TABLE 3.2.: Problem definition, involved actors, network structure, policy arena, and modes of interaction during the five phases of the process.

CHRONOLOGY	
Phase 1	The Onset of a Conflict: Disputes around the Global Abatement-Pollution Plan and First News about the Building of a Landfill (Feb. 1987-June 1988).

NETWORKS

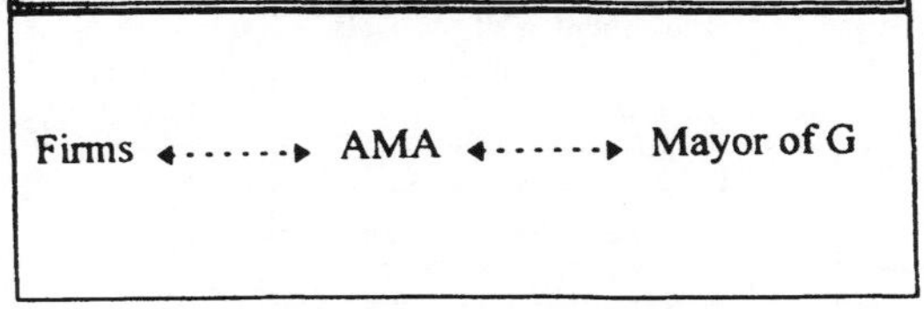

FIGURE 3.4. The network during the first phase of the process.

Phase 2	The Unleashing of a Conflict: Attempts at Building a Landfill and Social Opposition in Gibraleón (June 1988-Jan 1989).

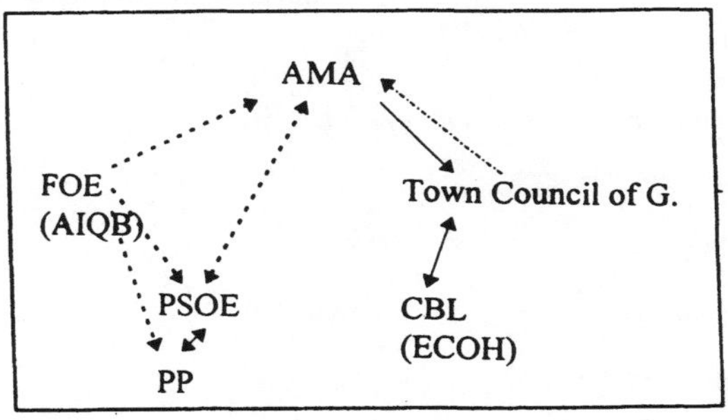

FIGURE 3.5. The network during the second phase of the process.

Phase 3	First Signs of Negotiation (the Setting of the Clean-Up Table) and the Shifting of a Conflict: The Substitution of Gibraléon by Palos de la Frontera as a New Site for a Waste Facility (Jan. 1989-July 1989).

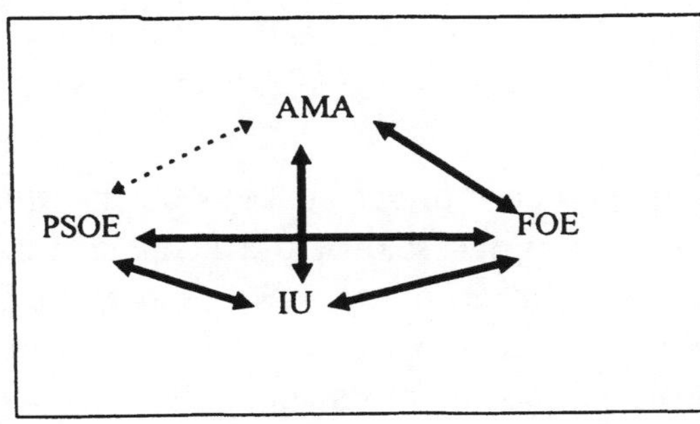

FIGURE 3.6. The network during the third phase of the process.

Phase 4	Initial Resistance against the Building of a Microencapsulating Plant in Palos de la Frontera (July 1989-Feb. 1990).

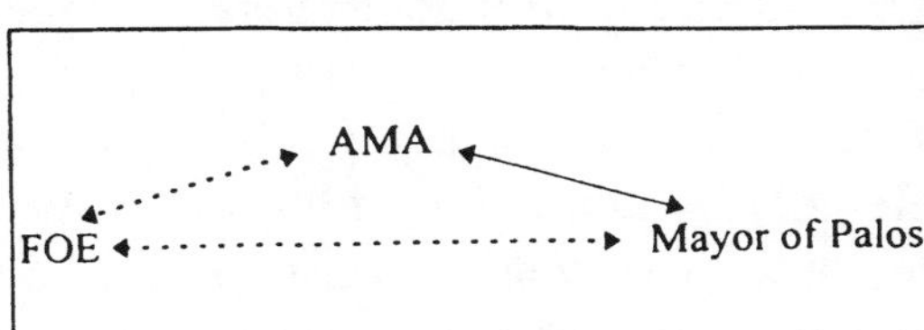

FIGURE 3.7. The network during the fourth phase of the process.

Phase 5	Final Solution of the Conflict: Vanishing Resistance and the Building of a Microencapsulating Plant in Palos de la Frontera (Feb 1990-1992).

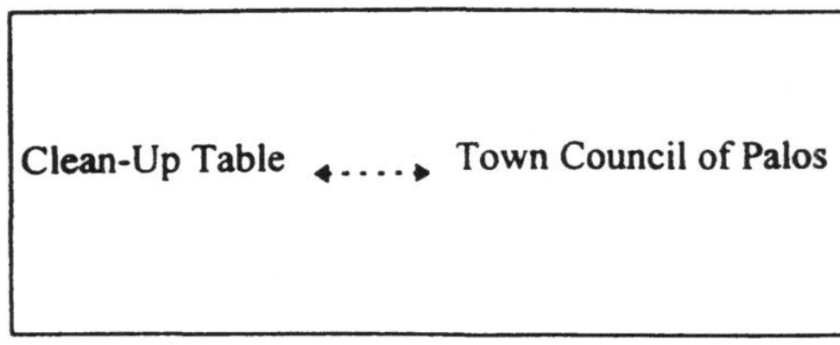

FIGURE 3.8. The network during the fifth phase of the process.

Explanation of symbols and lines:

Problem-solving

Bargaining

Conflict

Dubious stance

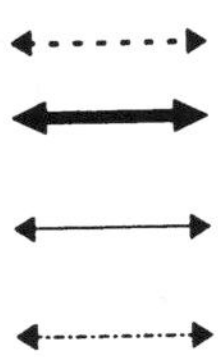

3.4 CONCLUSION: SUCCESS FACTORS

(A) The case under analysis clearly illustrates the importance of *social-learning* processes, or the fact that different actors (public and private) involved in common tasks tend to draw lessons from past experiences. However, in most cases the lessons drawn are not the same (the same collective actor might get different, and even divergent, consequences). Likewise, actors are not generally conscious of the existence of a learning process and consequently tend to proceed by means of a trial and error approach (that is, in a rather erratic way); the ingredients leading to success are not clearly rehearsed in successive experiences, nor are the elements which led to failures duly avoided. Social-learning operates in a much vaguer sense because actors are only aware of the need for not repeating past mistakes (they usually speak of 'before and after' a certain thorny issue or failure) but they do not know exactly how to avoid them. For this reason, a different method is favoured so that the assertive and, in many cases, impositional political style of some (public) actors lose weight to the benefit of a *piecemeal, gradual, and temptative approach* to the problem. This new approach does not naturally preclude the success of the process but it lessens, at least, the probabilities of failure. At the same time, there is no reason to conclude that the new learning will be definitively incorporated into the prevailing political style because, at least in the realm of waste facilities siting, the examples of failure are still more numerous than those of success. But failures, at least in some cases such as the one analysed in G., make actors aware of the need for reaching consensual solutions and thus create the basis for successes in the near future.

(B) In the case under analysis, a significant *process of institution-setting* has taken place. Two important institutions have been created as a result of the conflict unleashed by the failed attempt to build a landfill: The Clean-Up Table, a forum which, gathering public and private actors, forms a neutral arena in which to discuss a wide array of environmental problems in Huelva; and a Parliamentary Committee on Waste Management,[7] which tries to reach some consensus among political parties on the sites chosen for the construction of waste facilities. The Clean-Up Table has been the most important institution in the policy process since it contributed significantly to decreasing the degree of polarisation and confrontation which surrounded the landfill. The Table functioned as a political-institutional arena where conflicting interests gather with the aim of reaching consensual solutions. Although technical arguments were important in that forum, the Table succeeded in getting the different contending parties to agree on the need to bargain and respect certain rules of the game, such as to keep the discussions within the forum and not to take advantage of outside social unrest; to use arguments with the objective of reaching compromises, and to refrain from utilising the resources - unequally distributed - at their disposal in order to defend and implement specific and unilateral positions. This forum functioned, then, as a sort of political laboratory which left a blueprint on the actors; the people involved in environmental issues learnt that traditional approaches, based on imposition and criteria of expertise, were no longer valid. Besides, and more importantly, this successful experiment introduced a certain *degree of institutionalisation*

[7] The Parliamentary Committee on Waste Management was created in November 1992, after a conflict linked to a project for the building of an incinerating plant in Miramundos broke up. It embraces all the parties with representation in the Andalusian Parliament, and its president is a PSOE member.

and formalisation in the environmental policy process. There was also a process of *entitlement* whereby some actors were deemed legitimate to participate in the Table whereas others remained excluded. This change was significant because environmental policy, until then, had been mainly based on the political strength of the governing party and on the game of parliamentary majorities (it is important to bear in mind again that, in the Andalusian region, the PSOE has traditionally held the reins of power and has managed to concentrate resources of all types in its hands). The factor of unequal political strength undoubtedly contributed to a high degree of polarisation and to the extreme politicisation of the landfill issue: The main party in the opposition (PP) could use the strategy of siding with the opponents to the project to gain political power and legitimacy in the region. This strategy is, above all, reasonable if the opposing party has few chances to win at the polls; consequently, this is a pure partisan strategy which does not primarily reflect doubts about the content or substance of a particular project. The stance of the mayor of Palos, although belonging to the PP, is different however: in spite of the initial support for citizens' opposition to the micro-encapsulating plant, the content of the project is more carefully analysed and finally accepted - with some reluctance and not openly - by her. Besides, the fact that Palos is located within an industrial region and the mayor has personal links with industrial entrepreneurs, facilitate a political approach which is not exclusively partisan.

Although the two institutions created in the course of the process have been very favourably regarded, their results are somehow, and similarly, deceptive: they achieved the needed consensus among political forces (something that was missing in the mid 80s) but this did not dramatically transform the social stance towards waste management issues. Moreover, and more dangerously, if social consensus is not equally achieved (at least on a case-by-case basis), political consensus will turn out insufficient and politicians risk widening the gap between citizens' wishes and political accords. However, and despite this deficiency, the creation of the two institutions is proof of a certain learning process whereby political compromises should be reached - regardless of the distribution of political strength - before decisions on polemical issues are made.

(C) Another important element in the case of success in Palos is the approach of public agencies to the waste management issue: *the more consensual* (although restricted) *the approach* - generally accompanied by an information campaign - *the better*; by contrast, 'command-and control' approaches - only belatedly linked with information campaigns - tend to generate conflict. Yet, the convenience - or lack thereof - with respect to the setting of information campaigns is dubious; for instance, a good and 'unbiased' campaign can fail if citizens do not trust politicians to a large extent.[8] Despite this possibility, the probabilities of failure decrease if information is kept under control so that leakages to public opinion and attempts at manipulating pieces of news remain low.

(D) Along with the importance of piecemeal approaches and of institution-setting, one important point unravelled by the research is the *decline of experts' influence* upon controversial policy issues such as waste facilities siting. It has often been proved that experts can produce divergent documents, equally sustained by 'technical and objective' data, depending on the client. This fact would give *more room to political tools such as persuasion, reasoning, debates*, and emphasises therefore the importance of producing good *and credible arguments*.

(E) The *technical quality of the project* is undoubtedly connected with the probabilities of success as well. Although technical arguments are not the most important ones in controversial policy processes, the availability of an 'optimum' technical solution which has already been demonstrated and is correctly operating somewhere else contributes to the reaching of a solution. *'Arguments of comparison'* of the sort, 'Germany, a country with one of the most deeply sensitised societies about environmental protection, has long accepted this type of waste facility', can be useful in order to convince actors about the goodness of a certain project; if this argument is accompanied by a trip to the country and a visit to the plant in question, the chances of success are higher; the same would happen if someone connected with that foreign plant (preferably a non-suspicious figure such as a local, a farmer, an environmentalist) could travel to the host country and explain the qualities of the project. Yet this type of strategy is helpful if the political climate has not turned out very conflictual and polarised.

---

[8] On the whole, Spaniards trust their political incumbents less than the average European. In 1994 the Eurobarometer showed that Spain was, after Italy, the country in which the dissatisfaction with democracy ('I am not very satisfied or not at all satisfied with the way democracy works') was highest.

(F) The analysis has also led to discovering a connection between a more general acceptance of waste siting projects and the *industrial character of the village* in the vicinity. That is, the more the inhabitants of the place where the plant is going to be built are used to industry, the higher the probabilities of agreement on the project. If, on the contrary, the site is near rural areas with little contact with industrial processes, the probability that people will oppose waste facilities increases. This fact is related to different variables: the strategy of fear and the possibility of manipulating information on waste issues decrease if people are accustomed to the functioning of industries and to the nature of the firms' products. Besides, since people living near industrial areas may benefit from this activity in terms of jobs, compensation procedures (if the companies are labelled as hazardous) and the like, they will surely be more prone to accept waste facilities as an inevitable cost to be paid in exchange for the other benefits already received (and also as a necessary last step of the production process). On the contrary, if the village is relatively disconnected from industrial activities, the consideration of residues as something alien which is imposed against people's wishes increases.

Another important element is that, in towns located near industrial areas, the existence of *personal links* between the locals and the entrepreneurs and firms' staff tend to reinforce a positive view on the industry.

(G) The *type of estate* (public-owned and reserved for arable uses, or industrial uses; private-owned, etc.), where the waste facility should be located, influences administrative procedures and also the final result of the process. If the estate does not belong to the council, the proceedings are fewer and easier: The regional government can speed up the project and does not have to count on the council's approval. This is important because *the closer the administrative level to citizens' demands, the greater the possibility that the public actor in question* (the city council, the mayor, the councillors, the local party, etc.) *feels forced to side with its constituency*. This responsiveness, which basically appears in salient and controversial issues such as waste management, can easily lead to fights within the same party: the party which runs the council could oppose its party's policy in the regional government.

(H) The direct relationship between the salience of the issue under analysis and the quickness of citizens' mobilisation is also a point of interest. The traditional weak proclivity of Spaniards to undertake collective action to pursue common goals[9] and, accordingly, the preference shown for the setting of individual links with public authorities, seems to run counter to this finding. But no contradiction should be seen between the weak (comparatively speaking) system of interest intermediation prevailing in the country and the successful mobilisation of interests linked to single-issue movements. These social movements represent a good means for people who want to gain prestige and importance in small communities, and, at the same time, they are seen as the most effective voice procedure by citizens who are eager to express their views and concerns in the public sphere. This phenomenon may have some relationship with past experiences about the breaching of legal procedures, or with the implementation of polemical political decisions without previous consultation. For this reason, it would be better to *allow these social groups to follow a strategy of voice*, and to incorporate them, in a flexible way, into the decision-making process.

(I) Lastly, a situative variable such as the *economic crisis* can indirectly help to lessen the opposition to waste facilities. If industries are doing badly due to the economic situation, and, besides, they are forced to incur increasing costs because of a gradually stricter environmental policy, rumours about possible lay-offs and shut-downs will surely spread. The fear unleashed by the possibility of job losses could favour a more receptive social attitude concerning the need to build waste facilities as the only means at firms' disposal to lawfully eliminate their waste. The willingness to accept solutions to industrial residues will also increase if a *figure from the productive side* (leftist party representatives, mayors in industrial areas, unionists) *plays a predominant role in the decision-making process*. Nevertheless, the situation of economic crisis and the presence of relevant individuals from the productive side will only be effective in the case of sites near industrial areas.

[9] The following chart shows, in a comparative way, the low number of voluntary associations as well as the unimportant quantity of union membership in Spain (McDonough, Barnes, López Pina, 1984).

REFERENCES

Aguilar, S., Environmental monitoring and environmental information in Spain, in: Knoepfel, P., Weidner, H., Zieschank, R., *Environmental monitoring and reporting in selected countries*, Basel and Frankfurt: Helbing & Lichtenhahn, 1992

Hirschman, A.O., *Exit, Voice and Loyalty: responses to decline in firms, organisations and states*, Cambridge, Mass.: Harvard University Press, 1970

Junta de Andalucia, Agencia de Medio Ambiente, *Medio ambiente en Andalucia*, Informe 89, 1990

Ministerio de Obras Publicas y Transporte (MOPT), *Medio ambiente en Espana 1990*, Madrid: MOPT, 1991

Ministerio de Obras Publicas (MOPU), *DGMA: Actividades 1984*, Madrid: MOPU, 1985

LIST OF ACRONYMS

ADMINISTRATION.
- Junta: Regional Government.
- AMA: Andalusian Environmental Agency.
- MOPU: Ministry of Public Works at the central level (recently renamed MOPTMA, Ministry of Public Works, Transport and Environment).
- EGMASA: Environmental Management Company (public-owned).

PARTIES.
- PSOE: Spanish Worker Socialist Party (social-democratic oriented).
- PP: Popular Party (rightist moving towards the centre).
- IU: United Left (includes the Communist Party and some other minor political forces).

ENTREPRENEURS.
- FOE: Federation of Entrepreneurs of Huelva.
- AIQB: Association of Chemical Industries of Huelva.

UNIONS.
- UGT: Workers' General Union (socialist-oriented).
- CC.OO: Workers' Commissions (more communist-oriented).

ENVIRONMENTALISTS AND CITIZENS' INITIATIVE.
- ECOH: Ecologist Co-ordinating Organisation of Huelva (local environmentalist association).
- CBL: Co-ordinating Body against the Landfill of Gibraleón (ad hoc action committee).

4 ROLE OF A SCAPEGOAT NEEDED!: THE SITING OF A CHEMICAL WASTE INCINERATOR IN LENDAVA, SLOVENIA

ANDREJ KLEMENC, IGOR LUKŠIC

4.1 INTRODUCTION

In this chapter we analyse the decision making process concerning siting of the incinerator in LEK Lendava. The NIMBY syndrome emerged also in Slovenia and it was almost impossible to make a decision on waste facilities siting after the mid-eighties. LEK's story is therefore very interesting. Our analysis is based on the theoretical framework of policy analysis presented in this book. Some anthropological, discourse studies and risks society theory observations have been taken into consideration too. Data have been collected by twenty one interviews with actors and by analysis of the articles on the issue published in Slovene daily and regional newspapers as well as the minutes of important meetings. The main task of the analysis was to discover preconditions for the successful decision: successful in this case also meaning implemented.

4.2 EMERGENCE OF THE PROBLEM

In 1986, LEK[1], the large and successful Slovenian chemical and pharmaceutical company started the production of the antibiotic gentamycin in the industrial zone outside the town of Lendava on the Hungarian border. The transfer of gentamycin production to the underdeveloped, predominantly rural area of Pomurje was publicly opposed by the regional environmentalists, accusing LEK of transferring polluting technologies to the underdeveloped province of Pomurje.

[1]Pharmaceutical and Chemical Company, one of the largest enterprises in Slovenia, 2200 employees, 7 offices and 10 branches abroad; 1.3% of Slovenian GNP (1993); recently successfully completed a privatisation process.

B. Dente et al. (eds.), The Waste and the Backyard, 99–115.

Another trouble was that the legal environmental protection procedures had not been respected and the tests, proving that the emissions are in concordance with bye-law standards, were first carried out in 1992. But the communal government gave priority to jobs and tolerated the violation of procedures while the environmental grass roots activists were more occupied with issues of national importance, such as with spillages and an occasionally unpleasant and headache causing smell, provoked by decomposition of mycelia[2] - the by product of the production of antibiotics, stored within the Dolga vas Brickworks (BDV) before being added to the clay in production of the bricks. The complaints of the brickwork' s neighbours became a regular part of the agenda of the Local District Council (LDC), but both the representatives of the brickworks as well as those of LEK often took part in finding appropriate solutions to the problem.[3]

In the mid eighties, the region of Pomurje was one of the core areas of the Slovenian environmental movement, gaining the shape of an antipolitical opposition protest movement, widely supported by the media and the public. However, the main topic of the movement was to protect the picturesque Mura river from the expansionist plans of the Slovenian Power Utility. The success of the movement in protecting the Mura river has given a strong political identity and a corresponding power to the environmentalists in the region.

Nevertheless, by providing jobs and prospects for the development of the local economy and by a sensitive approach to the local community, LEK not only succeeded starting gentamycin production, but extended its production of antibiotics in 1992 and 1993. LEK, by providing about 300 jobs, became the only large employer with a good prospect of survival after Slovenia had proclaimed its independence and the Yugoslav market had burned out in the flames of war which followed. For being treated as 'domestic', also to the benefit of the local community oriented enterprise,

[2] Despite not being of any direct threat to the human health, mycelia had to be destroyed so as not to provoke resistance of the bacteria. Due to its high share of albumin substances it has to be incinerated in a very short time, because otherwise the unpleasant smell can cause lethargy and headache.

[3] To diminish complains of local inhabitants, deriving from periodic high emissions of soot and an unpleasant smell from BDV, the people from LEK and BDV often met with the Development Fund of the Republic of Slovenia representatives and together with the people of the Executive Council of the Communal Assembly discussed the problems in semiformal meetings in order to made improvements.

the business with the local brickworks turned out to be of great importance, beyond a pure commercial partnership. The brickworks was the oldest industrial enterprise of local origin in the town, centre of the local economy in the rural, underdeveloped area and the cultural and political centre of the Hungarian minority living in the region of Pomurje. Through maintaining the additional business and jobs by keeping the declining brickworks alive, LEK has lost the image of the foreign, 'imperialist' enterprise seeking only its own profits and thus gained a positive image within the local universe.

The war in former Yugoslavia has weakened not only the brickworks but also LEK's economic position. LEK made a decision to restructure its production and orient its export in a large part to the USA market, while Brickworks Lendava lacked the resources and capacities to face the new harsh reality with strategic restructuring. However, the birth of the new independent state of Slovenia not only provoked a negative economic impact on the local economy and dramatically worsened the traffic problem in Lendava, but also called into question the existing rules of the game within local politics, between the local and the state administration and challenged the established framework of the relations between the Hungarian minority and Slovenian majority at the local level.

Within the traditional, premodern political culture, the accumulation of the problems which could not have been solved either at the local level in the near future nor simply transferred to political powers on the higher level, the increase of frustration led to a search for a scapegoat. By acting with a sense due to the sensible use of its strategic knowledge about the other actors and the rules of the game, LEK escaped from receiving the status of the scapegoat in times when the whole community was looking for one - because of the accumulated problems and frustration.[4] LEK considerations were not restricted to the economic aspect of the waste business. After Brickworks Dolga vas was trapped in the scapegoat role,

[4]At that time the community was not only exposed to the invasion of international heavy truck traffic provoked by changed transport corridors because of the war in former Yugoslavia and had suffered from the decline of the regional economy, but was also frustrated because of the forthcoming new territorial and new administrative divisions brought into being by the reform of local self-management, which was also endangering the framework of the ethnic relations between the predominantly rural Hungarian local majority and the population of Slovene origin.

LEK faced no resistance in the implementation of the relocation and technological modernisation of its waste treatment.

In contrast to this, Brickworks Dolga vas made a strategic mistake by acting as an enterprise free from its incorporation in the community, although it was pressured by the needs of day to day survival and was exclusively oriented to following the immediate economic demands; it also made a mistake by overestimating the importance of the legal rules and neglecting the informal rules of local politics. The import of waste from Austria in 1993 thus turned out to be the lightning rod for accumulated social and political frustrations of the community, which resulted in victimisation of the director of the brickworks.[5] In the following dynamic, the media (re)constructed reality. LEK's and NIGRIN's waste became part of the same story, aiming to prove the inability of state powers to protect the people from environmental dangers, especially in the underdeveloped region, which is regarded by state authorities as a 'wasteland' province.

However, since LEK was able to separate its commitment to the local community from the servicing ('enbricking') of its waste in the local company exactly at the proper time, it was also able to present the relocation and modernisation of its waste treatment as a commitment to solve the environmental part of the problem. Consequently it has - in the shadow of the conflicts between the brickworks, the local community and the state - successfully implemented the solution according to the initial time schedule without facing the resistance of the residents and/or obstruction of the local authorities.

4.3 THE CHRONOLOGY

The policy process concerning the decision making about a (new) location of the LEK waste incinerator should be seen in four phases:

---

[5]Without being subjected to any legal charges he was trapped in a scapegoat role, to be sacrificed in a political ritual of the community, trying to survive its forthcoming institutional extinction and to maintain its power vis à vis the state, even after its own death as a political entity.

1. Decision on new incineration technology and new location
(Autumn 1991 - Spring 1992)
As a consequence of the war in former Yugoslavia, LEK in 1991 developed a new strategy, oriented primarily on the increase of sales of antibiotics and bulk chemicals to the USA and other Western markets. The decision to embark on the project of installing the incinerator within the LEK in Lendava, made in the autumn of 1991, was based on three presumptions: (1) Lendava Brickworks is not capable of surviving the economic recession; (2) the possible future demands of USA FDA Good Manufacturing Practice has to be anticipated; (3) The incinerator can improve the economics of production by providing the process heat.

In December of 1991 LEK requested the Institute for Environmental Protection in Maribor for an environmental impact assessment for the project and received a positive expert opinion in the spring of 1992.

It should be stressed that no other actors knew of that decision and the first steps of LEK.

2. The brickworks problems refocused the problem of LEK's waste
(Summer 1992 - Winter 1993)
In summer of 1992 the director of the Dolga vas Brickworks struck a deal with the Austrian firm NIGRIN. Neither LEK Lendava nor the local authorities, or The Development Fund of Republic of Slovenia - which had taken over the responsibility for the financial improvement of the brickworks - had been officially informed about the contract providing incineration of industrial waste from Austria. All the legal criteria for import and incineration of the waste had been respected; nevertheless the director and BDV were given roles of scapegoats for accumulated problems of the local community and also problems at the national level.[6]

[6] The waste policy is one of the weakest areas of the environmental protection in Slovenia, with quite a history of conflicts not only between the local residents and the authorities but also between the technical experts on the field, fighting on the definition of the best technical solution and neglecting the issue of social acceptance as of the secondary importance. In addition in 1992 two waste affairs broke out: the people in Zenica (Bosnia and Hercegovina) rejected the disposal of chemical waste from Slovenia; the waste has been sent back to Slovenia where no community was willing to accept the waste for temporary disposal. In the meantime the media gave attention to the statement of the director of one of the largest brickworks in Slovenia, accusing the brickworks in North Eastern Slovenia (Pomurje) of illegal competition by incinerating imported hazardous and toxic waste. He had BDV in mind.

Before the Christmas of 1992 the 'enbrickment' of mycelia in Dolga vas Brickworks stopped due to the insolvency of the enterprise. Since increasingly large amounts of mycelia were waiting to be processed in the Brickworks, the problem with the Brickwork's neighbours, blaming LEK Lendava for the trouble, started again.

At the beginning of 1993 LEK Lendava rejected the economically attractive offer to incinerate ('enbrick') the waste in brickworks of Ptuj. Rather then taking marginal and short-term profit and being blamed at the local level for breaking the rules of the 'agreement economy', LEK head office decided to invest approximately $90.000 in renovation of the local brickworks under the condition that The Development Fund of the Republic of Slovenia and local authorities would give guarantees for further 'enbrickment' of the waste until its own incinerator would start operating.

At the end of January 1993 LEK, for the first time, publicly announced its intention to install the incinerator in order to demonstrate its will to maintain its care for the community as well as its devotion to solving the environmental problems.

In February 1993 articles on the LEK Lendava waste problems were published in a popular regional newspaper, defining the production of antibiotics as a 'dirty technology transfer', as well as in large national daily newspapers, mentioning - predominantly in a negative context - LEK's decision to install the incinerator.

3. The *politicisation* of the waste treatment in brickworks

(March and April 1993)

In March 1993 the train with waste from Austria arrived in Lendava. Nobody was informed what was going on. In the absence of the director, the brickworks was not willing to give any information. The irritated residents informed the media and demanded a meeting with the director and the mayor. The latter was shocked because some months earlier the director of the state development fund gave him a promise that no waste business would be carried out to save the enterprise.

The regional, Pomurski vestnik, and the state media printed the stories, briefly reconstructing the whole history of the 'toxic waste' problem in Dolga vas Brickworks for the public. Within the next couple of days the regional radio station produced several talk shows about this the issue. From the perspective of the public it seemed that LEK and NIGRIN are the part of the same story: they are endangering local environment with incineration of the waste in Dolga vas Brickworks. The journalist claimed that the Basel Convention - not allowing export of hazardous waste - had been violated and that the business had also been bad in cash terms, since the costs in Austria would be 10 - 40 times higher than in the actual contract between NIGRIN and Dolga vas Brickworks.

The Head of Sanitary Inspection of the Republic of Slovenia, himself from Lendava, publicly expressed his dissatisfaction with weak control capacities of the inspection and outdated legislation on the field. He also disagreed with the extended time schedule for disposing of mycelia in the brickworks.

Although the director of Dolga vas Brickworks was trying to prove that up to the end of April 1993 legal procedures had been respected and that some other neighbouring brickworks had also been incinerating the very same waste before they went bankrupt in the recent past, and despite that later tests by The Institute for Environmental Protection in Maribor which proved the identity of the waste pattern, he was not able to legitimate the business to the local authorities and the public.

After the Italian firm SIATA had been chosen as a supplier of the incinerator and the project had been selected o n a tender of The Ministry of Environment of the Republic of Slovenia for the state supported implementation of the environmentally sound technologies, published in January, the Executive Board of LEK Holding confirmed the project in April 1993.

4. The ritual of scapegoat finding as conflict solution
(May 1993)
At the beginning of May 1993 at the joint session of all three chambers of the communal assembly of Lendava, after discovering that there was no legal background to the radical solutions, that is prohibition of incineration of any kind of waste in the Brickworks, the delegates recommend to BDV not to incinerate any waste at all.[7] The director of the brickworks and the state authorities in the field were sharply criticised.

Some days later the director of the brickworks in Dolga vas resigned, officially because of health problems.

In the following months the contract between NIGRIN and Dolga vas Brickworks was cancelled and the Brickworks was sold to a private firm after it had gone bankrupt in summer 1993. The imported wastes in Dolga vas are still waiting to face their as yet unknown fate.

In the meantime LEK has gained all permissions for starting the installation of the incinerator and gained the official license in summer 1994. In the beginning of June 1993, The Institute for Environmental Protection and Water Management of the Ministry of Environment and Regional Planning issued its Official Concordance. In July 1993 the Construction Permit for the construction of an incinerator in LEK Lendava was given by the Executive Council of the Communal Assembly of Lendava. The incinerator passed internal technical tests at the end of April 1994 and obtained, by decision of the State sanitary inspector, temporary permission for tentative operation for 3 months in the first week of May 1994. In the autumn of 1994, the Operating License was given.

4.4 TWO PROCESSES, NETWORKS, AND THE PATTERNS OF INTERACTION

In the reconstruction of the decision making process and political context of the installation of the incinerator in LEK Lendava the following two independent, but mutually interlinked processes emerged:

---

[7] Not taking into account the abstract and unspecified claims to abandon the industrial way of development and show more understanding for the greening of the local economy, made by the representatives of the Social Democrat Party of Slovenia, no specific party statements or claims have been made.

(1) the administrative process of the installation of the incinerator in LEK Lendava;
(2) the conflict on the actual waste incineration in Dolga vas Brickworks.

In our interpretation, the analysis of conflict dynamics of the latter is the key to explaining the success of the former, since only in this manner could the policy community, actors, their roles, networks and the rules of the game be discovered. The interconnection of the two processes has been found firstly in the investigation of the first one. Retrospectively, the first process turned out to be a part of the problem of the treatment of LEK Lendava's waste in BDV, which we shall attempt to prove in our interpretation.

The relations between actors in the two processes could be shown as follows:

Process I
Arena: Administrative and social

Actors	Problem definition	Objectives	Roles	Resources	Models of interaction	Results
LEK	Avoid resistance to the new technology; set up with (future) markets	Content	Promoter & Director	Scientific knowledge, strategy	Problem solving	Operating incinerator in time
Executive council and Communal Assembly EC & CA	Local employment	Content & process	Gatekeeper	Legal & political	Problem solving	Confirmed role
State sanitary inspector SSI	Promoting respect for environmental standards	Process	Gatekeeper	Scientific knowledge, legal	Problem solving	Confirmed role
Brickworks Dolga vas BDV	Economic survival employment	Content	Supporter of promoter as a scapegoat	Social	Bargaining	aggravation of image
Institute for environmental protection IEP	Promoting respect for environmental standards	Process	Gatekeeper	Scientific knowledge, legal	Problem solving	Confirmed role

Process II
Arena: socio-political and administrative

Actors	Problem definitions	Objectives	Roles	Resources	Models of interaction	Results
Brickworks Dolga vas BDV	Economic survival, employment	Content	Promoter	Legal, socio-political	Bargaining	Economic loss, damaged image, scapegoat status, director resigns
Local District Council LDC	Prevention of toxic waste incineration in BDV; solution of the LEK waste problem	Content	Opposer	Political	Conflict making	No additional waste incineration, enhanced role and position, improvement of image
Executive council and Communal Assembly EC & CA	Control keeping and peacekeeping consolidation of community	Process	Mediator, also filter	Political, interactive knowledge	Bargaining, conflict solving	Enhanced role, improvement of image, gain in prestige
LEK	Maintaining of waste incineration in BDV, not to be exposed	Content & process	No clear role	Strategy	Bargaining, problem solving	Shifting of opposed energy, assure waste servicing capacities
State sanitary inspector SSI	Lack of control on waste	Process	Gatekeeper	Legal	Problem solving	Enhanced role; improvement of image
Pomurski vestnik PV	Lack of the state control & responsibility for the regional development	Content	Opposer	Interactive knowledge, public opinion	Conflict supporting	Improvement of image

The first process should be seen as a purely administrative process which had begun with LEK's decision to construct an incinerator and ended by getting all needed documentation and permissions. LEK was director and organiser of the process. The procedure was very formal without any serious opposition.

The second process was crucial for understanding of the smooth path of the first one. The main actor was Brickwork Dolga vas, which tried to solve its problem of economic survival by incinerating mycelia from LEK. LEK was prepared for the game in which it gets the image of 'domestic' enterprise, nice for the local economy and jobs. In this process LEK played the role of being involved without any dirty interest of its own. In fact it canalised to BDV all the opposition energy of the local inhabitants, organised in a Local Community on the incineration of mycelia. It was a paradox that the inhabitants did not oppose the production and the producer of the mycelia, LEK, but the one who had mycelia as a productive source, i.e. BDV. Therefore Brickwork Dolga was recognised as a trouble maker by political will, i.e. the Communal Assembly of Lendava, and consequently this organ prohibited the 'enbricking' of mycelia. That solution would have allowed LEK to perform incineration in its backyard, as a trouble-free and technologically clean process.

Despite the fact that we are facing two separate (but very interfering) processes it is not only possible but in our opinion also wise not to take them separately while reconstructing the phases of the process in general. Not only because they are mutually interlinked and the majority of the core actors are present in both (with different roles, however) but also because we can confirm the well known finding that (in both processes) it is usually not the problem of the waste that is at stake, but the problem of the game between the actors. Precisely speaking, in our case the problem was not even in the game itself but in finding 'the name of the game' for the situation where the old local rules of a communitarian agreement economy had been broken because of the redefined state and constitutional meta-framework of the game (emergence of the new, independent, international market-oriented constitutional and political framework of the national economy on the edge of a military conflict). This has not only changed the structural position of promoter in process I by giving an advantage to the actors already integrated within the global market, but has treated the legal existence and 'socio-political' status of the local gatekeeper and the opposition. The fear of losing the legal and/or political possibility to participate in the game and thus no longer being able to legally influence the actions of the other actors has influenced both the opposition in process II as well as the (local) gatekeeper/mediator in process II. For the 'socialist self-management' decision making process the dominance of the 'socio-political' arena was characteristic. It was exactly the unique combination of having at the same time control over the legal resources of gatekeeper[8] and the political resources of the mediator, which had given the outstanding position to

[8] Including the political control over the gatekeeping resources of the state. The state inspection for example was institutionally integrated and thus subjected to the municipality as political entity.

the political representatives of the municipality within the arena. Since the legal competencies had been transferred over a wide area to the state authorities after the political and constitutional changes of 1990, while the local politicians remained responsible for the prospects of the local economy, they could not take a risk to invest the rest of the gatekeeping resources in order to keep their control over LEK, especially while LEK was strictly following the legal procedures. The purely administrative character of the arena within the process I is thus understandable. But this purely legal standing of the representatives of the community in process I has been compensated by emphatic use of the informal political power in process II, where a decision of 'no decision making' by municipal politicians was parallel to the political ritual of scapegoat finding, despite the fact that the promoter in process II (BDV) had also been acting strictly according to the legal procedures.[9] In this manner the purely administrative character of the process I and the overemphasised political character of the role of mediator of the process II has to be seen as referring to the same 'stake'. The actor to whom - due to the meta-contextual changes - a large part of the gatekeeping resources had been transferred (The State Sanitary Inspection) took the opportunity to play a political game in process II, aimed at the process content, to balance its de jure status with its de facto role. The domination of the traditional community politics over the party politicisation of the conflict has further influenced the network structure and the patterns of interaction in both processes. The role of the political opponent, which first made the conflict within the community possible, has thus been played by the local media - a very common situation in the post-communist transition period. This political role of the media has also influenced the behaviour of the second opposition - BDV - by encouraging him to act publicly and not to relay on personal contacts with other actors.

[9] One cannot neglect two important aspects which are relevant, beyond the large difference in economic resources of both promoters. LEK's standing as a private entrepreneur has been in concordance with its legal status of privatised stock company, while privatist behaviour of the director of BDV was not in concordance with the legal and de facto status of the firm. Secondly, while private company LEK kept informing the local leaders about its intentions, the business with 'Austrian waste' seemed to be hidden even from the state restructuring agency as de facto owner of the company.

4.5 CONCLUSION

Considering the situation variables that have in this case contributed to success we could emphasise:
(1) the worsening of the local economy due to decline of the local industry;
(2) the outbreak of the new, larger problem - the expansion of international lorry traffic through the village - by its nature more attractive to raising the claims of the community toward the national political and institutional arena;
(3) the tradition of institutionalisation of policy making within the political culture of 'self managing corporatism' in Slovenia, enabling the strong interaction between the socio-political and socio-economic arena.

In this respect it seems that the dominance of pro-development policy orientation and the level of the institutionalisation of policy making are also in this case the important elements of success. However, we can in our case hardly speak about *political* pro-development orientation, since at the level of politics on the industrialisation based, political discourse has lost its hegemony and has been replaced with 'small is beautiful' discourse, promoting the development of traditional agriculture, tourism, crafts and small industries. Because the latter discourse was only articulated in opposition to the communist industrialisation ideology and not as a distinctive and articulated political program, it was not able to challenge the pro-development image of the promoter. Rather than domination of one political orientation we have been facing a lack of the political and the domination of pragmatic communitarian discourse. With respect to the industrial culture we can also confirm its importance in the Slovenian case; however, we again have to emphasise its peculiarity.[10]

[10]It is definitely not the industrial culture as a set of values which give the central position to industrial labour and an outstanding role to the industrial workers, emphasising their industrial collective identity, the institutions of collective representation and bargaining with the representatives of the employers and the state. Within the area dominated by small scale authentic farming, the industry gives additional income at the individual level, social and health insurance through the employment of one or more members of the family, whereas on the collective level it contributes to the development of the local infrastructure. The industry is still considered as a community servicing activity, run and controlled by the leaders of the community. From the other side, the industrialisation of small scale farming has contributed to the distribution of certain technical knowledge, which has given roots to the industrial technical culture among the farmers.

This social background can in our opinion contribute to explaining the nature of the opposition in process II. This was neither a semi-formal citizens' group, nor a local NGO, nor a political party. LDC was representative of the local district, institutionally incorporated within the socio-political arena of the community. It was not a representative of the community (i.e. the statements, interests and opinions of the residents of the district) because it was the community itself.

The following factors in the process of decision making contributed in our opinion to its success:

1. Selected distribution of information
In process I, LEK temporally and spatially selectively distributed information about its intention to construct an incinerator. After the decision to start the project had been made, the information was distributed to the most influential local leaders through informal channels and only after the problem of 'enbrickment' of waste in BDV in the beginning of 1993 had become serious and had LEK Lendava already transferred the responsibility of 'enbrickment' in Brickwork Lendava to the local government and the state restructuring fund, was the information presented to the public. Contrary to this, it was, in our opinion, the hidden character of the BDV business with NIGRIN that contributed to the escalation of the conflict. By not informing the top local politicians about the waste business, the director of the Brickworks was endangering the very structure of the power of local authorities since this was already endangered by the forthcoming reforms of local self management.

2. The strategic presentation of the new technology as a solution and elimination of the old problem
Instead of a promotion campaign based on scientific knowledge, which also turned to be contraproductive in Slovenia, LEK simply presented the incinerator as a better solution to the emerging incapability of Brickwork Lendava dealing with waste without breaking its positive image as creditor of the local economy. It was not the technological superiority of the new technology as such that was emphasised, but the elimination of the transport of waste and relocation of the incineration within the factory

which, in contrast to BDV, was situated in an industrial zone, far from settled areas.

3. Respect for the rules of the game
Although the renewal investment of BDV was in fact an advance payment for future servicing of its waste, it also contributed to maintaining LEK's image as an enterprise that is committed to the community. By the nature of its origin and the scientific character of its production perceived as 'foreigner' (or 'outcomer'), LEK, by co-operating with BDV, also symbolically enforced its image of 'domestic' enterprise, operating in favour of and not against or at the expense of the community. If LEK would only optimise its own option by transferring the waste to the other temporary incinerator, the positive-sum game framework between LEK (benefiting from disposing of waste), BL (having a strong business partner and creditor) and ECCG (feeling responsible for local employment and emphasising its intermediary role within the community) would disintegrate. In this case, ECCG as the mediator could use its political and cognitive resources to redefine LEK's positive symbolic image within the community by using its legal resources (of postponing and temporising the siting and construction permissions) as gatekeeper in the Process I. LEK succeeded in leaving the game without breaking the contextual framework it benefited from. In fact it was BDV who broke the rules of the game within the policy community by importing waste from Austria and thus endangering LEK by overloading its 'enbrickment' capacities and by giving the attribute of 'foreignness' to the waste in general.

4. The production of a role of the scapegoat in Process II
Although BDV did not break the law it first broke the rules of the game by not informing EC&CA about the business with NIGRIN. The role of EC&CA as mediator was thus not respected; it was even more endangered in time when the future of the design of the political and institutional arena was already uncertain. By not protecting BDV with claims for the respect of the rule of law (there could not be any doubt about the legality of the waste business), BDV - at that time 'shared' by the local community (EC&CA) and the state development found - was symbolically expelled from the community and served as a scapegoat, enabling EC&CA to design a conflict as a symbolic conflict between the

foreign and domestic, state and community, abstract law and agreement economy. The resignation of the director of BDV, cancellation of the contract with NIGRIN and the political consensus not to incinerate waste in BDV any longer, could be considered as a redistribution of costs and benefits, resulting from the zero-sum game started by BDV. Since the costs and benefits have been (and would also remain by the operating incinerator) balanced and LEK succeeded in carrying on its 'domestic' image, LEK was silently excluded from suffering from the consequences of the political result of the process II, which thus only aimed at foreign waste. In fact the very installation of the incinerator could be retrospectively interpreted as a contribution to the implementation of the political will not to incinerate waste in BDV any longer. Therefore we can point out subsequent factors of success.

5. The mediator did not strive to maintain and institutionalise its position after the solution was found

In our case the situation in this particular point is very special. The 'political' actors (EC&CA, LDC) were in fact the actors that are as such vanishing from the scene because of the ongoing process of reforming the territories, structures and competencies of the local communities. Thus the personal dimension of politics has been emphasised from the one side, while from the other the community vs. state symbolism of the conflict expressed the endeavours of symbolic unification of the actors and powers in the community in order to compensate the rapidly coming diminution of legal and administrative competencies. In that sense the process II could also be interpreted as a ritual performed to keep the power of the community alive after the death of the institutional actors and their networks.

The key role in the explanation of success is played by the strategic approach of LEK in dealing with the other actors and the community in general. LEK was in fact the only actor who actually developed a strategy and a reflective attitude towards other actors as well as towards its own activities and actions. Being aware of its position, how it is perceived by the others and the position of the others, LEK was able to learn quickly from the conflict dynamics and to effectively use interactive knowledge as its main resource. This was due to the fact that LEK was well anchored in the community - the management of LEK Lendava comes from the region and is very experienced in local politics (the former director of

LEK Lendava and LEK's representative on the Executive Board of BDV, after the latter was taken over by Development Fund of the Republic of Slovenia, was President of the local Government at the beginning of the 80s, etc.). Contrary to the other actors, LEK as promoter also had at his disposal time and money resources to turn interactive knowledge into strategy, which was based on a comparative cost/benefits approach, evaluating options not only by the direct/benefits monetarised equivalents, but also respecting the contextual long-term impact on its own position in the game. The priority of the jobs to the green anti-industrialist ideology in the period of economic recession and the focusing of the local media and public consciousness on the increasing road and transport problem, facilitated the implementation of the strategy.

5 SWISS BORDER INCIDENT: A CASE STUDY OF THE CIBA-GEIGY SPECIAL WASTE INCINERATION PLANT IN BASEL, SWITZERLAND

URS ZUPPINGER, PETER KNOEPFEL

5.1 INTRODUCTION

Ciba-Geigy (hereafter CG) is one of three major pharmaceutical companies located in the Basel area; the other two are Sandoz and Hoffmann-La Roche[1]. Due to its geographical situation, the Basel area has an influence which spreads beyond the borders of both canton and country. As a result, these companies are a driving force of the local economy and employment market.

In 1984, the management of CG decided to incinerate its special waste instead of having it treated in managed special waste disposal sites. To this end, the company drew up a project to build a special waste incineration plant (SWIP) in Basel-Stadt. The plant was also to treat the special waste delivered by outside clients.

This project was greeted with approval by the other local pharmaceutical companies (Sandoz, Hoffmann-La Roche) and by the cantonal authorities of Basel-Stadt and Basel-Land. The future plant was seen as a project of regional scope.

Nevertheless, CG wished to retain full ownership of the plant as well as full control over all operations.

CG is located in the Klybeck and Schweizerhalle production areas, in Basel. The SWIP is located on the Klybeck site, in Kleinhüningen[2]. This area is located in the upper northern part of the Basel-Stadt canton. Near the industrial plants and the CG headquarters, there is a residential neighbourhood. Further to the north and the east lies the German city of Weil am Rhein, in the district of Lörrach, Baden-Württenberg. To the west, across the Rhine, are the French municipalities of Saint-Louis and Huningue. The Schweizerhalle site, which CG shares

---

1 CG and Sandoz have now merged under the new name of 'Novartis'.
2 See figure 1.

B. Dente et al. (eds.), The Waste and the Backyard, 117–159.

with Sandoz, among others, is situated in the Basel-Land canton, to the west of the city of Basel, near the Rhine.

On the basis of experiments conducted in the 1970s, the management of CG concluded that treating special waste in managed waste disposal sites was increasingly costly and difficult. Furthermore, CG had acquired a great deal of competence in the field of waste incineration and the company's high level of technological expertise in dealing efficiently with pollution problems was already widely recognised.

In 1972, on the Klybeck/Kleinhüningen production site, not far from the site of the future SWIP plant, CG installed the K224 rotary furnace which could incinerate 3,500 tonnes of waste per year. Four other furnaces, partially integrated into the production processes exist on the company's production sites in Monthey (in the canton of Valais), and in Schweizerhalle. The Schweizerhalle furnace could burn 9,000 tonnes of waste per year and the steam thus generated was used in the production process.

In 1983, CG offered Hoffmann-La Roche to incinerate the dioxin-contaminated waste from Seveso in the K 224 furnace in Basel. After careful preparation, 41 barrels — 2.5 tonnes — of dioxin-contaminated waste were successfully incinerated in the spring of 1985. The event was monitored by an external neutral organisation and covered by the media. CG thus proved its high level of competence in the field of special waste incineration.

In 1984, CG decided to build a new plant in Klybeck, capable of treating 12 to 15,000 tonnes of waste per year. The plant would be linked to the railway and road networks and to the company's internal road network. The buildings would be no larger than the surrounding constructions.

The decision-making process for the new SWIP took place in a changing legal context, just after the implementation of the new Law on the Protection of the Environment (hereafter LPE).

The new law states the 'obligation to recycle, neutralise or eliminate waste' (article 30, title). The role of the federal state is to enact directives (art. 32) and ensure co-ordination (art 31.4) The cantons must execute the law, either on their own behalf or by delegating responsibility to local authorities or private enterprises (art. 31).

The LPE was voted in by the Federal Chambers on October 7, 1983 and came into effect on January 1, 1985. Between 1986 and 1990, a series of enforcement ordinances followed, among which the Clean Air Ordinance, implemented on March 1, 1986, the Ordinance on Movements of Special Waste, implemented in April 1984, the Ordinance on Impact Studies (January 1, 1989), on Waste Treatment (February 1, 1991) and on Protection against Major Accidents (April 1, 1991). The entire decision-making process of the SWIP, including the procedure, surveys, discussions and conflicts, took place in the context of the new law and of the debates that preceded the enforcement of several ordinances, in particular the impact study ordinance and the waste treatment ordinance.

This process was also affected by the local population's loss of confidence in the Basel pharmaceutical industry after a storage warehouse for chemicals, on one of Sandoz SA's production sites, burned down in November 1986. The new SWIP project met with heightened opposition in Basel-Stadt and in neighbouring German cities, resulting in a permanent public debate on the issue. After the appeal to the competent cantonal court failed, however, the opponents to the project gave up trying to settle matters through court and began to negotiate directly with CG.

Our case study follows the sequence of events and procedures as recounted by the press and in interviews by those who were most closely involved in the process. The study is also based on a number of documents we were able to obtain[3] during our research. The case history can be divided into three phases :

Phase 1 From the official announcement of the project to the Schweizerhalle fire.

Phase 2 From the application for a preliminary permit to the granting of the building permit

Phase 3 From the registration of the appeal to the laying of the foundation stone.

---

[3] For a list of these documents, see the publication 'Usine d'incinération de déchets spéciaux (UIDS) Ciba-Geigy à Bâle (Special waste incineration plant (SWIP) Ciba-Geigy in Basel)', Urs Zuppinger/Peter Knoepfel, September 1995, IDHEAP, Lausanne.

FIGURE 5.1 The site of the sepial waste incineration plan in Klybeck, seen in its regional context

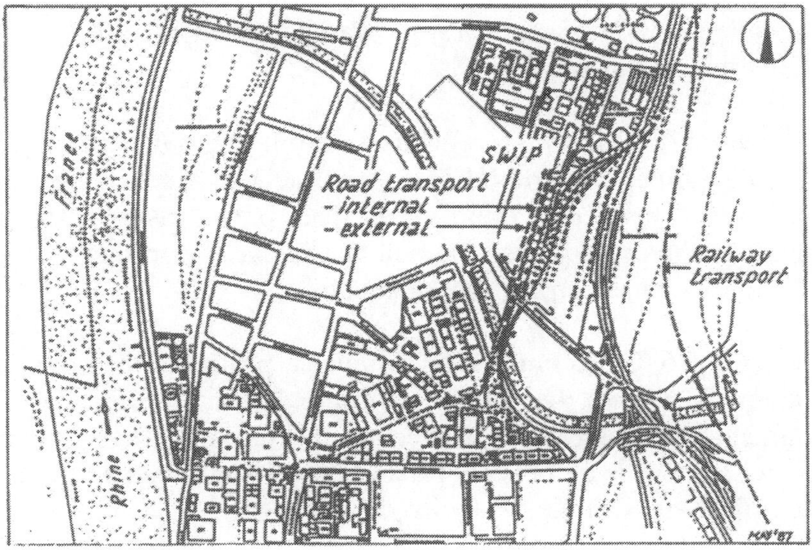

FIGURE 5.2 Plan of the site

5.2 CHRONOLOGY OF EVENTS

Phase 1 : From the official announcement of the project to the Schweizerhalle fire

CG announced its intention to build a SWIP on the Klybeck production site at Kleinhüningen during the first semester of 1985. The aim of the plant was to meet the needs of CG and of two other pharmaceutical companies, Sandoz and Hoffmann-La Roche.

On August 19, 1985, a delegation from CG informed the Commission for the Protection of the Environment of the cantonal parliament[4] of the Basel-Stadt canton (hereafter CPE/CP) of the project. The canton authorities then inquired whether it would be possible to incinerate the special waste collected by the public services in these facilities. CG accepted. The neighbouring canton of Basel-Land also showed an interest in having the waste collected by its own services incinerated in the CG facilities.

4 'Grosser Rat' : the legislative body of the canton.

On December 23, 1985, the project was approved in principle by the State Government[5] of Basel-Stadt.

On February 14, 1986, CG organised an information session for the residents of the neighbourhood concerned at 'The Three Kings Inn'[6] in Kleinhüningen, near the future site. The press covered the event. Opponents to the project later claimed that invitations to this session had been sent only to a selected group of people.

On April 4, 1986, CG submitted a preliminary project to the CPE/CP. Dr. B. Böhlen, head of the Federal Office for the Protection of the Environment (hereafter FOPE) attended the meeting. He approved the CG project and expressed his confidence in the company's ability to carry out the project and operate the facility.

Following this meeting, several members of the commission (belonging to different political groups) submitted a written request to the State Government, demanding that an independent commission be created to monitor the project. CG approved this request on September 16, 1986. On October 28, 1986, the State Government approved the creation of a monitoring commission staffed by specialists from both the public and the private sectors. This commission would be in charge of ensuring that the future SWIP was run according to regulations and of submitting a yearly report to the State Government. This report would be included in the yearly report submitted by the State Government to the cantonal parliament.

In June 1986, the FOPE published the 'Guidelines for waste management in Switzerland'. According to these guidelines :

- Switzerland should make greater efforts to destroy its own waste;
- as far as technically possible, waste should not be stored for long periods of time but treated immediately, so that resulting residues may be stored definitively.

---

5 'Regierungsrat' : the executive body of the canton.

6 'Gasthaus zu den Drei Königen'

That very same month, a group of residents of Kleinhüningen announced the creation of the group SMOG[7] to the press and launched an information campaign in the neighbourhood. Their arguments were the following :

- incineration technology is as yet imperfect;
- security measures in case of accident are insufficient and the neighbourhood will suffer;
- the transport of special waste in trucks is dangerous.

However, the SMOG group had difficulty spreading its views.

On November 1st, 1986, a serious accident occurred in Schweizerhalle, on a Sandoz production site, when a warehouse storing chemicals burned down. The fire was intense, with explosions (according to the newspapers, the flames reached 50 m). An orange cloud smelling of rotten eggs spread over the city, causing throat and eye irritations. The disaster alarm was sounded. Since the water used to quench the fire was drained into the Rhine, the river was heavily polluted and underwater fauna was destroyed over dozens of kilometers downstream from Basel.
In Basel, an organisation called 'Aktion Selbstschutz'[8] was created, and its membership rapidly rose to 1,500. A permanent secretariat was set up and a periodical, the 'Gasette', was published under the responsibility of Mr. H.G. Heimann, a SMOG activist. SMOG joined 'Aktion Selbstschutz' as an independent working group, while pursuing its own action separately.

In Weil am Rhein, just over the German border, a local action group by the name of 'Morgenluft'[9] was created by Gabriele and Johannes Foege, lawyers and environmental activists.

---

7 'Sondermüllofengegner(-innen) : opponents to the special waste furnace.
8 Action Self-Defense
9 Morning Air

Phase 2 : From the application for a preliminary authorization to the granting of the building permit

On February 16, 1987, despite the uproar caused by the Schweizerhalle fire, CG submitted an application to the cantonal administration of Basel-Stadt for a preliminary building permit. The contents of the application were rather 'light' : the information provided described only the physical characteristics of the facility (location, size, access) and its main activities :

- incineration capacity : 13,500 tonnes/year (2.4 t/hour).
- type of waste : solid, liquid, viscous
- amount of slag produced : 1,350 tonnes/year, to be delivered to the Elbisgraben inert materials refuse dump, in the canton of Basel-Land.
- estimated cost of the project : 100 million SF[10], (including 25 million for smoke-washing).
- cost of incineration : from 300 to 1,000 SF/tonne.

Concerning environmental questions, an 'emissions declaration' was included in the application.

The application was submitted to a public enquiry to last 30 days as of February 26, 1987.

This official confirmation of the project's existence, barely three months after the Schweizerhalle fire, triggered the anger of opposition groups. The protest movement thus gained in strength and audience, and improved its organisational structure :

- Among the groups opposing the project during the public enquiry were 'Morgenluft', SMOG, 'Aktion Selbstschutz' as well as the 'Democratic lawyers' and the Socialist party of the Basel-Stadt canton.
- The German action group 'Morgenluft' established contacts with SMOG.

---

10 Swiss francs

- The SMOG group aired its views during a press conference on March 22, 1987. Its arguments were based on principle : to accept the project, with all the risk and uncertainties involved for the population of Basel, would mean recognising that the actual level of production of special waste is inevitable. Thus, before even solving the problem of special waste disposal, it is important to obtain guarantees that the production of such waste will be reduced in the future.
- The German lawyer H.P. Schmidt launched an information campaign among the residents of Weil am Rhein. The information was transmitted to the local legislative body. The Weil am Rhein municipal authorities and the district office of Lörrach expressed their concern.
- In Freiburg im Breisgau, the president of the Land of Baden-Württenberg, Dr. N. Nothelfer (Christian Democrat) requested that the application for a permit and the surveys be submitted to a panel of German experts in charge of evaluating the impact of the project on the environment in the area situated in Germany near the site. In Dr. Nothelfer's opinion, the Swiss and German authorities' ability to reach an agreement in this case represented a test of their ability to co-operate in general. In May 1987, the German request was rejected by the Government Councillor E. Keller, head of the Public Works Department of the Basel-Stadt canton (hereafter DPW/BS).
- In November 1987, the SMOG group conducted a survey in the residential neighbourhoods of Kleinhüningen by asking residents to fill out forms concerning their perception of foul smells[11] in the area. The aim of the survey was to heighten the population's awareness of the unpleasant smells coming from the nearby production site.

In the other camp, the project's sponsors and supporters also hardened their positions :

- The head of the DPW/BS announced that CG would have to provide a environmental impact report (EIR), in accordance with the LPE

[11] 'Stinkblätter'.

and the recommendations made by the FOPE while preparing the OEIS[12].

- In April and May 1987, several members of CG management publicly hinted at the possibility of moving the Basel production site to another location if the SWIP project was rejected.
- On May 25, 1987, CG organised a joint press conference with the heads of the DPW/BS and the FOPE, in order to make it clear that if the project was not accepted, the Federal Government could force the canton of Basel-Stadt to build a SWIP.

In September 1987, the project received the conditional support of a German official institution, the committee of the association of German municipalities of the Hochrhein-Bodensee region. The conditions were the following : the project's impact on the environment had to be acceptable, and CG had to agree to burn the special waste of the German municipalities.

During the summer of 1987, the media blew both hot and cold on the issue :

- On June 4, 1987, CG presented its project in a television programme on special waste, filmed in CG premises. Following opponents' protests, Mr. Heimann of the SMOG group was invited to participate in the programme. 'Aktion Selbstschutz' organised a street demonstration. The protesters marched into the Klybeck site and dispersed after a window was broken in the room where the programme was being filmed. The incident did not lead to any serious consequences, and no other demonstrations of this type were organised.
- On July 8, 1987, Martin Matter, a journalist from the 'Basler Zeitung'[13], published an article on the project. His arguments were the following :

  After the Schweizerhalle accident of November 11, 1986, economic and employment reasons could no longer be accepted as sufficient justification for authorising the construction of the SWIP.

---

12 OEIS: Ordinance on environmental impact studies.

13 The main daily newspaper of the canton.

Building a SWIP near the special waste's production site cannot be avoided. Creating such a facility, however unpleasant it may be, is in keeping with the rules of good management. Therefore, to be acceptable, the project must be irreproachable. The successful incineration of the dioxin-contaminated waste from Seveso cannot be considered a sufficient argument in favour of the project, since the new facility's scope will be much greater.

While the public debate raged on, meanwhile, in Basel-Stadt and in Bern, administrative and technical procedures were moving along. At the cantonal level, the case was taken over by the Interdepartmental Commission for the Protection of the Environment (hereafter ICPE), which brings together representatives of all the competent structures of the cantonal administration.

- In April 1987, the ICPE (canton) and the FOPE (federal administration) together with CG, prepared the brief[14] for the environmental impact report (EIR).
- In June 1987, CG entrusted the EIR to Suiselectra, a Basel-based consultant firm.
- In 1987, after evaluation of the state's share of responsibility in the Schweizerhalle incident, the cantonal authorities of Basel-Stadt decided to set up a 'Co-ordination Office for the Protection of the Environment'[15] (COPE), headed by a new recruit, Mr. A. Schrade, a lawyer who had previously worked at the FOPE and participated in the elaboration of the Law on the Protection of the Environment (LPE). As of January 1, 1986, this group was in charge of encouraging and co-ordinating the efforts of the canton in environmental matters. The Interdepartmental Commission for the Protection of the Environment (ICPE) continued to exist, but it was now led by the political head of the DPW/BS, instead of one of the department directors, as had previously been the case.

The brief of the EIR was sent for consultation to various environmental organisations, which in turn hired the services of consulting firms to examine all the legal and technical aspects of the question.

---

14 'Pflichtenheft'; today this brief is request by the OEIS

15 'Koordinationsstelle für Umweltschutz'.

In a similar vein, the authorities of Weil am Rhein requested the services of experts to assess the legal ways and means of intervening in the project, as well as the level of risk for the population. The experts were a chemist and an engineer from Freiburg im Breisgau and two legal specialists, Reinhard Zweidler of Basel and Hans-Peter Schmidt of Freiburg im Breisgau. The latter, as the reader may recall, had launched an information campaign concerning the SWIP project among the residents of the city in February 1987. The authorities of Weil am Rhein later entrusted both lawyers with the task of defending their interests.

The experts' report, which cost 16,520 DM, was published in December 1987. According to its conclusions, the future SWIP would increase the pollution levels of an already badly polluted area. Using the smoke-washing technology would be risky, given the lack of experience. The report criticised the fact that if ever the system were to malfunction, the gases produced by the furnace would be directly let out into the atmosphere through an emergency chimney. The authors also insisted that maximum security measures should be ensured for the storage of special waste before incineration.

In February 1988, CG filed an application with the Basel-Stadt authorities for a building permit. The application included the EIR. This document provided most of the technical aspects of the project, since CG's application did not include any technical details.

The project was then submitted to a public enquiry to last 30 days as of February 17, 1988.

The project's opponents on both side of the Swiss-German border decided to co-operate. These protest groups included SMOG, the WWF, 'Aktion Selbstschutz', the SPD and the 'Morgenluft' local action group of Weil am Rhein, as well as the 'Velo' association from Lörrach and the 'Lörracher Bund'. SMOG called upon the population to protest against the project during the public enquiry period. With the help of 'Aktion Selbstschutz', the group distributed a leaflet to the residents explaining how to express their disapproval of the project, and set up a special secretariat to help them.

On March 14, 1988, the 'Morgenluft' local action group invited the management of CG to the presentation of an 'expert survey'. Two hundred people attended the event. Mr. A. Schaerli, head of production at the Basel CG factory, was present at the meeting, along with twelve specialists. They managed to provide satisfactory answers to all the participants' questions, except for those concerning the issue of dioxin formation, which they admitted required additional study.

According to the press, by the end of the public enquiry, 451 protests had been officially registered. According to Mr. Schrade, 800 individual persons or collectives had expressed disapproval; in fact, these protests were contained in 95 separate documents totalling 800 pages. The protests came from both Germany and Switzerland.

In a press-conference, the Basel and Swiss WWF organisations, 'Aktion Selbstschutz' and SMOG, along with three scientists (a zoologist, a toxicologist and a chemist) and two jurists declared the following :

- First, they believed that the SWIP should only be built within the framework of a national strategy to eliminate the production of special waste. Such a strategy was to be urgently defined.
- Secondly, a list of all the 'defects and inconsistencies' of the project was presented as proof that the project was not viable.
- Lastly, the organisers of the press conference provided a long list of additional information and documents still lacking; they demanded that the authorities establish a reference document concerning the level of pollution in the area before the construction of the SWIP. An adequate monitoring process, they said, should include the right for opponents to intervene during the elaboration and preparation of the project; furthermore, the building permit should be delivered only under certain conditions.

The cantonal administration began to feel that they could not cope with such an uproar, given the legal and technical complexity of the issue. Following the advice of the COPE, the Zürich Ökoscience institute[16] was hired to do the following research :

16 Institut für praxisorientierte Ökologie (institute for practice-oriented ecology).

(a) an independent evaluation of the quality of CG's application for a building permit and of the EIR;

(b) a systematic evaluation of the arguments and requirements voiced by the EIR's opponents.

On June 17, 1988, Ökoscience handed in two reports which were transmitted for evaluation to the COPE and the ICPE. The latter concluded that the EIR was inadequate and, at the COPE's suggestion, drew up a list of the additional information to be provided by CG.

On July 5, 1988, SMOG embarrassed the project's promoters by publishing excerpts from a report by Suiselectra entitled 'Report on Waste' which had not been publicized during the public enquiry. The report indicated that the amount of special waste produced in the area was higher than expected and that a second furnace had to be built[17]. According to the project's opponents, the authorities' claim that the non-publication of the document was allowed under the LPE was not admissible, since some of the information included in the report was considered, even by the press, as being critical of the way waste was being managed by the local public services. Mr. Schrade of the COPE admitted to having been mistaken and announced the publication of the document.

On September 20th, 1988, the canton authorities told CG that the EIR needed to be further documented. The deadline for turning in the report was set to January 31, 1989. The additional information concerned the following questions:

- the waste flow (Stofffluss), from the delivery of the waste to the cleaning and elimination of residues,
- operating the furnace,
- cleaning the air,
- protecting the water,
- protection against noise,
- risk situations.

---

17 Indeed in November 1988, CG applied for a building permit to extend its Schweizerhalle rotary furnace for special waste. The permit was granted, and the facility's capacity increased from 15,000 to 24,000 t/year (according to Spiegel, 21/90). This permit triggered only a single protest which was registered only after the end of the public enquiry period... This lack of interest in the issue deserves to be reflected upon.

The German press then published articles stating that the system CG was planning to use no longer corresponded to present needs and that its technology was now outdated (global incineration, without any differentiation between the different types of chemicals), but that CG was going to stick to it for economic reasons.

The authorities of Weil am Rhein, believing that the additional demands made by the cantonal authorities to CG on September 20th, 1988, were insufficient, filed an appeal with the cantonal Commission of Appeals for matters concerning Building Regulations of Basel-Stadt [18] (hereafter CABR).

In an interview published in the October 1988 issue of the 'Gasette', Mr. Schrade explained the importance and the limits of the Environmental Impact Report. Since the facilities were privately owned, the owner company, according to the LPE, had no obligation to prove that the project was needed. The company's obligation was to prove that everything has been done to ensure that the impact of the facility on the environment met the legal standards. If not, the authorities could close down the facility. Given the nature of the legal context, the co-ordinator of the Basel authorities advocated the setting up of a permanent monitoring commission. He also expressed his personal opinion that the results of the monitoring must be publicised and proposed that a public hearing be organised to promote dialogue with the project's opponents.

In January 1989, Mr. H.G. Heimann of SMOG explained in an article published in the leftist newspaper WOZ[19] that detailed and constant criticism of the project would force the producers and managers of special waste to innovate. This position was a slightly less radical than the opinions previously expressed by the group.

On March 15th, 1989, CG completed the requested additional report[20] for the EIR (400 pages), concerning the 52 points which needed clarification according to the canton authorities' decision of September 20, 1988.

---

18 Basler Baurekurskommission.

19 'Wochenzeitung' ('Weekly')

20 The initial deadline of January 1 had been extended three and a half months.

According to Mr. A. Schrade (COPE) and Mr. A. Schaerli (CG), this document was the result of multiple contacts and working sessions between CG and the cantonal administration. Besides yielding concrete results, this collaboration also created an atmosphere of mutual trust between the two parties.

According to the press, the report stated the following :

- An analysis of the waste flow (Stofffluss) made it possible to determine the maximum amount of different types of special waste which could be incinerated without exceeding pollution limits. The study also showed that dioxin or furan could not be detected either in the slag or in the water flows or in the expelled air; concerning the effectiveness of the nitrogen oxide elimination system, a security risk factor would be taken into account; however, federal standards would be respected. The best way to eliminate heavy metals still remained to be determined.
- Responsibilities would be specified in a precise description of the monitoring procedure for the special waste delivered to the site.
- The report also included an evaluation of possible accidents.

The report was published and copies were sold for 100 SF[21]. Contrary to what had been announced in the press in September 1988, the project, with the new, improved report, was not resubmitted to public enquiry.

The German public authorities then decided to enter the debate. On March 25, 1989, on behalf of the municipal authorities of Weil am Rhein, the community of municipal administrations of the 'vorderes Kandertal' and the city of Grenzach-Wyhlen, the lawyers Zweidler and Schmidt announced that the waste flow analysis was incoherent and contained false information.

In Basel, however, an atmosphere of conciliation seemed to prevail. In early April 1989, an article by the journalist Martin Matter, published in the Basler Zeitung, described the project in detail intimating that the results now inspired confidence.

---

21 Swiss francs

The lawyers Zweidler and Schmidt, in their written depositions to the cantonal authorities on behalf of Weil am Rhein and the Basel opponents, criticised the additional report and proposed that the building permit be delivered only under certain conditions. These conditions were not only technical; thus, according to their claim, CG should be under the obligation to reduce its waste production levels, the building permit should be granted with a time limit, under the condition that the facility be updated according to the regulations of the moment and the technology available at the time the permit expires. The lawyers' criticism of the additional report was published in the June 1989 issue of the 'Gasette'.

In June 1989, an article published in the German press stated that opposition to the project was waning in Basel and that the German municipalities were now leading the movement. Furthermore, according to the article, the opponents were now no longer demanding that the project be abandoned but were only trying to ensure the best possible conditions for its implementation. The local action group 'Morgenluft' denied this and confirmed its total opposition to the project.

On June 19, 1989, Mr. Schrade of the COPE, supported by the ICPE, the State Government and CG, organised a public hearing on the SWIP in the Basel Fair building, which lasted more than five hours. The hearing brought together representatives of the federal and cantonal administrations, independent experts, a delegation from CG and a delegation of opponents, with their respective experts, representatives of the local authorities, members of parliament and journalists, all in all about two hundred people. The themes discussed were :

- the quality of special waste produced in Switzerland today and in the future;
- the efforts made to reduce the amount of waste produced;
- the choice of technology and whether or not to build large-scale, non-selective facilities;
- the problem of dioxin and the necessity of establishing standards.
- the feasibility of the smoke-washing system for the elimination of heavy metals and salts.

On almost all points, the Swiss and German opponents disagreed with CG and the representatives of the Swiss cantonal and federal administrations. Mr. A. Schrade declared to the press that the hearing had made it possible to pinpoint the aspects needing further research in order to frame the conditions for the building permit.

At the end of November 1989, the COPE completed the environmental impact report on the future SWIP, which it elaborated with the agreement of the ICPE. According to this report, the Basel authorities had requested clarifications which were not included in the additional impact report, the reference text for the COPE report. The latter concluded that CG's answers were satisfactory concerning water protection and emergency measures in case of accident. The documents still lacking could be provided at a later stage, and would then receive administrative approval. In conclusion, the authors of the report said that the permit could be delivered under certain additional conditions, of which a long and detailed list was included.

The report also advocated the setting up of a monitoring commission made up of representatives of the public authorities on one hand and of an advisory board on the other. The advisory board would in turn be made up of representatives of the company in charge of operating the furnace, the furnace's users, the local authorities and the local residents, as well as representatives of regional environmental organisations. The number and names of associations included was still to be determined, but they would be limited to those based in Basel-Stadt.

On December 8, 1989, though it was already being rumoured that the decision was imminent, 'Morgenluft' organised a public protest meeting against the SWIP, and invited doctors to speak. The meeting was well attended (60 people). Leaflets were handed out.

On December 12, 1989, the canton granted CG the building permit. The required conditions pertained to building regulations. The COPE report dated November 30, 1989 was declared an integral part of the decision-making process to grant the permit. The FOEFL[22] and the State

---

[22] The FOPE is now called FOEFL — Federal Office of the Environment, Forests and Landscape

Government of the Basel-Stadt canton publicly expressed their satisfaction.

CG announced in a press communiqué that it agreed to fulfil the conditions set by the building permit decision.

The German opponents, represented by the lawyer H.R. Schmidt, declared that postponing important issues until subsequent administrative authorisations was illegal. The authorities of the Basel-Stadt canton, for their part, denied this and asserted that the permit could be granted since the reports and analyses had all shown that all the conditions guaranteeing the protection of the environment would be met. The rest was only a question of technical adjustments which could be monitored through administrative means. The real impacts would be monitored once the facility became operational. If these impacts did not meet the legal standards, the state would then impose additional requirements.

Phase 3 : From the filing of the appeal up to the laying of the foundation stone

After the decision, the project's opponents nevertheless persevered, especially in Germany. The main instigators were the authorities of Weil am Rhein, pressured by their voters/residents whose awareness of the issue had been heightened through the efforts of the 'Morgenluft' local action group. In retaliation, the Basel politicians who supported the project tried to jeopardise the collaboration between Basel-Stadt and Weil am Rhein.

In December 1989, 57 appeals were registered before the expiration of the deadline. In fact, the actual number of appeals formulated in writing amounted to twelve.

In January 1990, the press publicised the opposition of the German cities (Weil am Rhein, Lörrach, Grenzach-Whylen, the community of local authorities 'Vorderes Kandertal').

The appeals were submitted to the CABR[23], a commission presided over by the head of the DPW/BS, Eugen Keller, an avowed advocate of the project (see Phase 1). The commission appointed two experts who requested additional technical information from CG. These steps were taken without the appellants' knowledge.

On January 5, 1990, Morgenluft organised a public meeting in Weil am Rhein. The local authorities were present and 250 people attended the meeting. CG and the Swiss cantonal authorities refused to participate. The meeting confirmed the opposition to the project and the authorities of Weil am Rhein were asked to continue the protest.

In the March 1990 issue of the 'Gasette', Morgenluft explained its standpoint; the arguments of the association centred mainly on the problem of dioxin.

In April of the same year, the German press reported the change of position of the president of the Land of Baden-Württenberg. According to the newspaper, Dr. N. Nothelfer, who in 1987 had been one of the project's first opponents, had written to the head of the DPW/BS requesting permission to send German waste to Basel.

In July, the project's proponents began to react. On July 20th 1990, the deputy Gallacchi, vice-president of the Basel-Stadt Christian Democrat party, a chemist and staff member of CG-Basel's security department, supported by 79 other deputies, proposed before the cantonal parliament of Basel-Stadt that the agreement between Basel and Weil am Rhein allowing the latter to send its household waste to the Basel cantonal incinerator be rescinded. Mr. Gallacchi made no secret of the fact that this motion was intended as retaliation after the appeal filed by the authorities of Weil am Rhein against the SWIP.

In August 1990, the opponents received the expert report ordered by the appeals commission. They were given two weeks to decide on their position.

---

23 Commission of Appeals for matters concerning Building Regulations of the Basel-Stadt canton

The CABR held a meeting on September 5, 1990, which lasted 8 hours. Both parties (represented by R. Zweidler for the opponents and L. Müller for CG) were heard. The experts provided answers to the 40 additional questions submitted by the opponents after reading the expert report.

On October 19, 1990, the CABR rejected the appeal to the 'decision to grant a building permit' dated December 12, 1989; the motives of certain appellants were declared inadmissible (Lörrach), because the latter were located more than 2 km away from the site (this decision was made public only on November 29, 1990). CG, however, agreed to fulfil the additional conditions concerning the cleaning of water and promised to build an additional water treatment plant if it turned out, once the incinerator was operating, that the heavy metals produced by the washing of the smoke could not be eliminated in the existing water treatment plants.

The CABR's decision was made public during the fall session of the Basel-Stadt cantonal parliament. During this session, two-thirds of the parliament members supported Mr. Gallacchi and asked the parliament to study the possibility of rescinding the agreement allowing the city of Weil am Rhein to send its household waste to the Basel incineration plant.

On October 25th, 1990, the mayor of Weil am Rhein, Dr. P. Willmann, member of the German Christian-democrat party, was invited to participate at an open round table of the Swiss Christian-Democrat Party on the theme : 'The protection of the environment on a world-wide scale'. Among the persons present at this meeting were Mr. F. Cotti, member of the Swiss CDP and head of the federal department in charge of the protection of the environment, Mr.E. Keller, member of the Swiss CDP and head of the DPW/BS and Mr. J.J. Weber, member of the 'French CDP' and president of the General Council[24] of the Haut Rhin department. Dr. P. Willmann had already had several discussions beforehand with the German CDP and with Dr. Böhlen, member of the Swiss CDP and head of the FOEFL.

[24] 'Conseil Général'

In October 1990, the mayor of Weil am Rhein received the answer from the German Office for the Protection of the Environment[25] in Berlin, to which he had submitted the SWIP case. The Office's response was rather positive, with nevertheless a few restrictions. Several questions deserving further clarification were also pointed out.

The authorities of Weil am Rhein then decided to modify their approach. On December 13, 1990, the mayor met with the management of CG in order to begin direct negotiations, though he also declared that legal action had not yet been abandoned.

The first meeting was held in the presence of Mr. Randegger, Mr. Krauer (general managers of CG), Mr. Schaerli (co-ordinator of the project on behalf of the public authorities), Mr. Müller (CG lawyer) and Mr. Naegelin (technical director of the project). The lawyers R. Zweidler and H.P. Schmidt were also present. According to the mayor of Weil am Rhein, the furnace did not meet all the German standards concerning emissions. CG promised to comply and confirmed this commitment in writing. The statement was written by the CG lawyer, Mr. Müller, and its contents were discussed with the mayor of Weil am Rhein.

However, in spite of these concessions, new appeals were filed on December 28, 1990, on behalf of several German municipalities, among which Weil am Rhein, and of several Basel environmental protection organisations (the WWF among them).

On January 24, 1991, the legislative council of Weil am Rhein officially acknowledged CG's statement which followed the negotiations with the mayor, but decided to postpone its decision concerning the new appeals. The council commissioned the local administration to negotiate with the cantonal authorities of Basel-Stadt.

On February 19, 1991, the legislative council was informed that during discussions with the Basel authorities and with CG, the latter had provided additional guarantees concerning the monitoring of waste before incineration and had committed itself to providing the concerned German towns with maximum information on the programmes and the results of

25 Umweltbundesamt

the measures taken, as well as on the administrative authorisations to be delivered at a later date. The legislative council then decided to withdraw the appeals.

The lawyer R. Zweidler upheld the appeals on behalf of the WWF/BS and other Basel organisations. The only organisation still paying him fees was the WWF.

During the next month, discussions took place between the WWF and the Basel-Stadt local authorities.

Following these discussions, on April 19, 1991, in a new written statement, CG promised to respect the 'target levels'[26] of emission, which were much lower than the legal standard of particles and nitrogen oxide, and to respect an additional 'target level' of dioxin emission, for which no standards were specified in the Swiss legislation. CG agreed to let the SMOG group sit on the future advisory board, promised that the latter would be informed of the results of all measures taken by the company and granted the WWF the right to evaluate the technical basis of the incineration and smoke-washing facilities.

Having received these guarantees (which had no legal value), the WWF/BS, 'Aktion Selbstschutz' and SMOG withdrew their appeal. The only appeals which remained pending were those filed by a German fishers' association and by Mr and Mrs Foege, the leaders of the 'Morgenluft' action group. These appeals were also withdrawn several days later. According to the two German lawyers, the appeals were withdrawn not because they had obtained satisfaction, but to avoid running the risk of setting a legal precedent, since the decision was likely to be negative, the opponents remaining a small minority.
The legal battle thus ended without the case being sent to the Federal Court[27], a procedure which would have delayed construction yet again.

The foundation stone was laid on December 7, 1992, in the presence of the management of CG and of the new head of the FOEFL, Philippe

---

26 CG avoids mentioning the word 'standard'. The difference between 'target level' and standard reflects the company's 'philosophy of action', which differs from the usual practices and customs.

27 The final appeals court in Switzerland.

Roch, formerly president of the Swiss WWF. The facility is now completed and running. Public debates have ceased to focus on this issue.

5.3 THE NETWORK OF ACTORS AND ORGANISATIONS

Phase 1 : From the official announcement of the project to the Schweizerhalle fire

This phase is divided into two stages (before and after the Schweizerhalle fire).

Stage 1.1. The official announcement of the project.

At first, the majority of actors were favourable to the project, and their aims and actions for the most part converged (see fig. 3) :

- Most actors chose to sponsor or co-promote the project of CG.
- CG took on the role of project leader. In the early eighties, the company realised that new ways of eliminating special waste had to be found. By that time, CG had become very competent in the technique of incinerating this type of waste, as was demonstrated by the highly publicised incineration of the Seveso waste. Furthermore, after the adoption of the LPE in October 1983 by the Federal Chambers, CG knew that environmental restrictions would only become more stringent. The new general manager, Mr. Krauer, was also intent on promote his company's image. His aim was to show that CG was open to dialogue and aware of the importance of protecting the environment.

  Practically speaking, the convergence of many interests made the project particularly appealing. Indeed, a normal-sized furnace could incinerate much more waste than that produced by CG alone; the other local pharmaceutical companies (Sandoz, Hoffmann-La Roche) had similar needs but did not have CG's technological competence. On the part of the cantons of Basel-Stadt and Basel-Land, there was an even greater demand (and lack of technological competence) for the elimination of the special waste produced by

- The FOPE, which in Switzerland was in charge of planning and controlling the elimination of waste, was from the start in favour of the project. Indeed, such a plant could be made into a model facility.
- The issues raised by the project were discussed at the cantonal parliament of Basel-Stadt; the conflicts were resolved when all the parties involved accepted the principle of creating a technical commission in charge of monitoring the facility.

The only sign of protest on the part of the population came from residents living near Kleinhüningen, supported by the environmental activists of the SMOG group. The aims of the project, which at first were generally accepted, were the following :

- to eliminate the special waste produced by the Basel pharmaceutical industry.
- to eliminate the special waste produced by households and by other local companies;
- to respect the environment in compliance with the LPE implemented in the spring of 1985.

If this combination of favourable elements had been maintained throughout the process, the project could have been rapidly completed, and the debate would have only concerned the solving of a technical problem at minor cost and in compliance with the legislation on the protection of the environment.

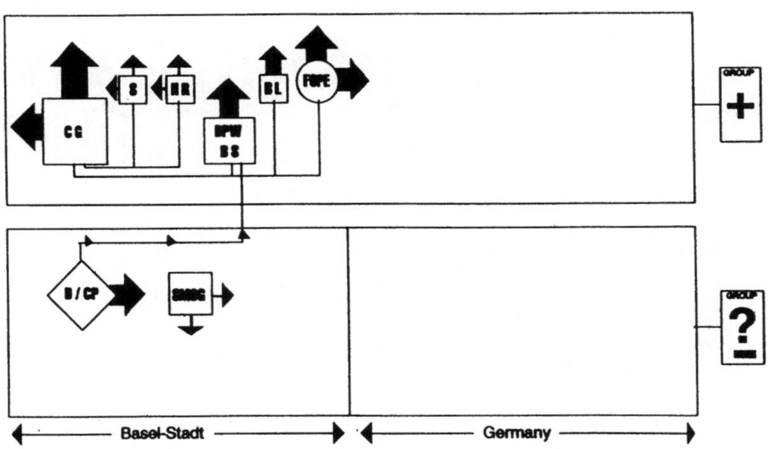

FIGURE 5.3 Network of actors. Stage 1.1. Before the Schweizerhalle fire (for a key to the abbreviations see the list at the end of the document)

Stage 1.2. The Schweizerhalle fire

After the Schweizerhall fire, several protest groups emerged in Basel-Stadt and in Germany, in Weil am Rhein ('Aktion Selbstschutz' and 'Morgenluft'). The SMOG group also intensified its action (see Fig. 4).

After the fire, the population lost confidence in the local pharmaceutical industry. The management of CG was thus confronted with the necessity of pursuing a new, more global and political aim : the SWIP was to be a way of regaining the trust of the citizens of Basel. CG then set up a 'non technical project management' group, a small working group directly linked to the company's top management. Public relations were entrusted to Mr. A. Schaerli, head of the Klybeck production site, who had considerable experience in matters of communication, knew the company's interests and relied on qualified staff in charge of the technical design of the project. He was all the more able to devote himself entirely to the project since he was about to retire in the summer of 1988.

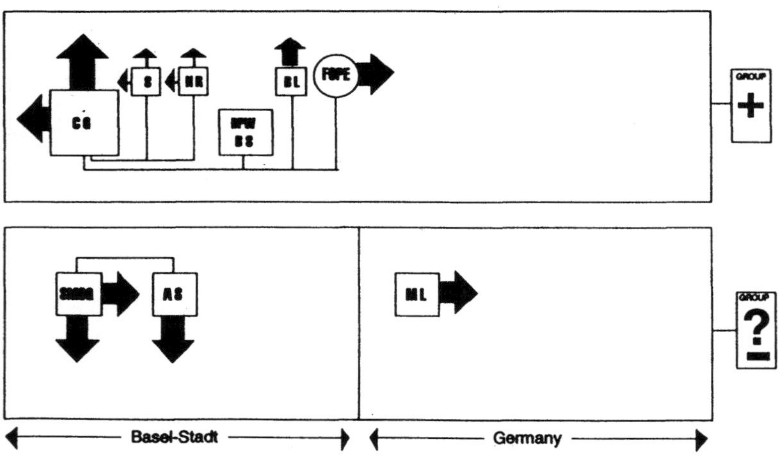

FIGURE 5.4 Network of actors. Stage 1.2. In the wake of the Schweizerhalle fire (for a key to the abbreviations see the list at the end of the document)

Phase 2 : From the application for preliminary authorisation to the granting of the building permit.

Stage 2.1. 1987 : crystallisation of the proponents' and opponents' positions; the search for the right strategy.

In the spring of 1987, aware of the importance of the issues raised by the project, the Schweizerhalle fire and the new legal context, CG decided to initiate the first official procedures (the application for a preliminary authorisation). This step modified the network of actors (see Fig. 5).

Once the project was opened to public enquiry, the Basel protesters increased their action, more in reaction to the Schweizerhalle fire than to the SWIP project itself. However, throughout the entire process, the opposition garnered little political support. On the German side, the project's opponents managed to involve both local (Weil am Rhein), and regional official representatives and organisations (Lörrach district office; Dr. N. Nothelfer, president of the Land of Baden-Württenberg). The different groups of opponents established links, including cross-border links; nevertheless, the opposition remained divided :

- 'Aktion Selbstschutz' took a radical stand ('Better unemployed than a victim of chemical pollution')[28], but the SWIP was not the central focus of its action.
- The WWF supported the opposition but remained discreet. The political parties moves seemed dictated mainly by political considerations.
- In Basel, the SMOG group was the only actor whose action was exclusively focused on the SWIP project, which it was ready to accept under the specific condition that a general policy aimed at reducing the amount of special waste were adopted.
- In Weil am Rhein, the 'Morgenluft' action group adopted a radical stance to defend local interests. The local residents were mobilised to pressure the authorities of Weil am Rhein. The latter were not opposed to the project but requested guarantees concerning the quality of the project. The other German municipalities remained hesitant.

---

[28] in German : 'Besser arbeitslos als chemietot'.

- The president of the Land of Baden-Württenberg became involved in the case because of the fact that interregional co-operation was being challenged rather than strictly because of the SWIP issue.

Despite the Schweizerhalle accident, the authorities of Basel-Stadt maintained their position; however, they tried to play a specific role in the debate :

- the political authorities of Basel-Stadt, along with the FOPE, publicly supported CG and maintained a hard-line position (CG threatened to delocalise jobs; the authorities announced that the Federal Government could impose the SWIP),
- the administration of Basel-Stadt saw to it that the LPE was abided by. Thus, the 'declaration of emission' submitted by CG for the preliminary application had to be replaced by an EIR, according to the law. CG agreed to provide this document. The cantonal administration requested that CG participate in drawing up the brief of the report along with the FOPE. CG entrusted the EIR to an outside consulting firm, Suiselectra.

The Basel authorities then made a fateful decision: to organise a consultation with various environmental organisations, concerning the brief of the report.

These organisations turned to consulting firms for help on the technical and legal aspects of the document. In August 1987, the authorities of Weil am Rhein also ordered a technical and legal expert assessment.

It seems that due to the Basel authorities' decision, the various protest groups were forced to enter the battle on technical and legal grounds. However, though these groups' arguments against the project were increasingly pertinent, they were also increasingly difficult to understand for the population. Furthermore, while these organisations' resources were necessarily limited, the services of consulting firms were, on the other hand, very expensive.

In the summer of 1987, the media gained interest in the issue. A television programme on the SWIP was filmed in the CG plant, with the participation of the leader of the SMOG group. During the filming,

'Aktion Selbstschutz' organised a demonstration which ended in broken glass and total confusion. After this failure, 'Aktion Selbstschutz', whose attitude was rather radical, chose to remain in the background. The Basel opposition groups were from then on dominated by SMOG, whose principal aim was to heighten the awareness of the residents of the Kleinhüningen neighbourhood and to use all possible legal avenues, as well as the media, to attain his goal. The debate was widely covered by the press on both sides of the German-Swiss border, but the television seemed to avoid the subject.

The Basler Zeitung criticised the hard-line position adopted by the CG — DPW/BS — FOPE coalition. In an article published on July 8, 1986, the journalist M. Matter declared that though the SWIP was indeed necessary, there was a price to pay for the Schweizerhalle accident, which had revealed the deficiencies in public control over such facilities.

The main local newspaper thus pressured the authorities into taking an interest in the quality of the project.

It seems that the political authorities did not ignore these demands. Until the end of phase three, political support for the project was never unconditional. The Schweizerhalle experience led the Basel authorities to set up a 'Co-ordination office for the protection of the environment' (COPE), and appoint Mr. A. Schrade, a jurist, as its director. Mr. Schrade, a former member of the FOPE, had worked on the elaboration of the LPE. He was now in charge, with the help of others, of co-ordinating the SWIP case. As both 'gatekeeper' of the project (in charge of verifying legal aspects) and mediator between the parties, Mr. Schrade played an important role in resolving the conflict. Mr. Schrade, who fully agreed with the journalist M. Matter's position, took advantage of the pressure exerted by the project's opponents and tried to improve the project through negotiation and by using all possible legal avenues.

Both then and later, the authorities of Basel-Land did not intervene in the debate. However, their position always matched that of Basel-Stadt. The FOPE's attitude was determined by the fact that the federal administration had faith both in the Basel authorities' ability to keep the project under control and in CG's technical competence. The federal authorities had an interest in the successful outcome of the project :

a national policy for the protection of the environment was being elaborated and CG's SWIP project was considered a first step towards the domestic elimination of the total amount of special waste produced in Switzerland.

The application procedure for a preliminary authorisation gave CG a chance to assess the situation. On the one hand, it appeared that despite the Schweizerhalle incident, the opponents' position against the project was not very strong. On the other hand, the local authorities had to be stringent in matters of environmental protection and could not express too openly their support for the project. These considerations were taken into account by CG to determine its course of action.

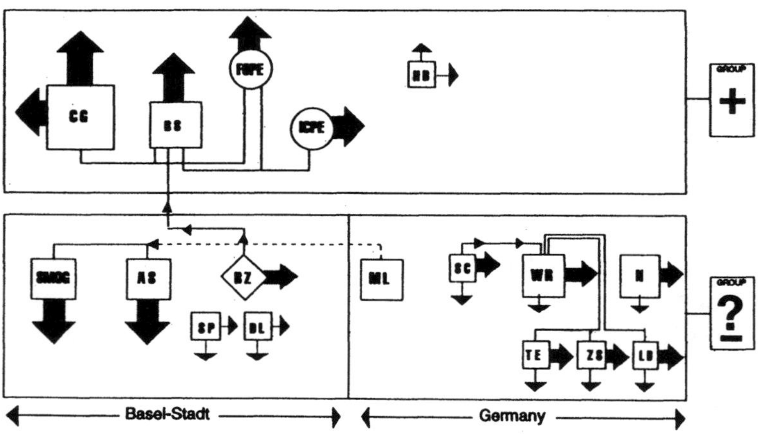

FIGURE 5.5 Network of actors. Stage 2.1. Request for a preliminary authorisation, 1987 (for a key to the abbreviations see the list at the end of the document)

Stage 2.2 Serious negotiations precede the granting of the building permit.

The application for a building permit, which did not contain an elaborated technical project, was submitted to public enquiry, along with the EIR. As a result, the different parties' roles were modified.

In both countries, protest groups began to act. Co-operation between non-institutional actors of both countries improved. SMOG encouraged local residents to organise protest movements. CG accepted 'Morgenlufts' invitation to participate in an 'assessment survey' and recognised the need for an additional enquiry concerning the risk of dioxin formation during the smoke cooling phase. The press gave detailed explanations of the opponents' arguments, which were well researched and unconventional.

The cantonal authorities accepted the challenge and had the EIR evaluated by an external consulting firm. The latter's conclusion that the EIR should be improved was taken into account. The authorities drew up a long list of additional information to be provided and requested an additional impact report from CG.

When SMOG revealed that part of the document prepared by Suiselectra hadn't been made public, Mr. Schrade immediately complied to these demands and published the missing part. In his evaluation of the project, Mr. Schrade also drew inspiration from some of the opponents' proposals (for example, the suggestion to grant the building permit under conditions), while ensuring their compatibility with the administrative procedures in progress and with the law. In an interview with the opponents' 'Gasette', he explained the legal framework of the canton's action and proposed the creation of a permanent monitoring commission whose members would include representatives of the opponent groups.

At the same time, Mr. Schrade took advantage of the time needed by CG to prepare the additional report, to establish an atmosphere of trust between the COPE and the technical and non-technical staff in charge of the SWIP project at CG. The final points of the additional report were negotiated in Mr. Schrade's office.

After the additional report was completed, the Swiss opposition seemed to lose steam. The Basler Zeitung journalist, M. Matter, wrote that the results of the additional report inspired confidence.

In Germany, on the other hand, the local authorities increased their action. Weil am Rhein was no longer alone in the battle. The city was represented by two lawyers, Mr. Zweidler and Mr. Schmidt; however, the latters' aims did not exactly coincide with those of their clients. Indeed, as the LPE's enforcement ordinances were being prepared, their aim was also to ensure that the law's procedure and quality requirements were made more stringent.

In June 1989, Mr. Schrade organised a public hearing which confirmed in the eyes of the public that the negotiations between the State and CG were yielding positive results.

The building permit, granted on December 12, 1989, took the COPE report into account. Though this report was even stricter than the additional report in terms of requirements, it nevertheless allowed for several points to be controlled subsequentiy by means of supplementary administrative authorisations which could not be subject to appeal.

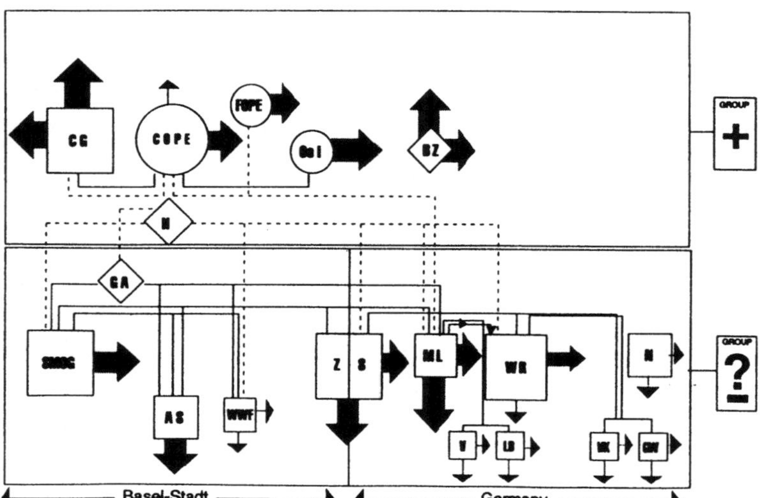

FIGURE 5.6 Network of actors. Stage 2.2. Application process and granting of the building permit, 1988 - 89 (for a key to the abbreviations see the list at the end of the document)

Phase 3 From the registration of the appeal to the laying of the foundation stone.

Stage 3.1. The appeals; pressure is exerted to speed up the procedure.

Once the building permit was granted, the project's proponents tried to avoid having the project sent to the Federal Court, since such a procedure would again delay construction. The strategy they adopted was the following (see Fig. 7) :

- CG and the cantonal authorities ceased all dialogue with the opponents. The COPE and its head, Mr. Schrade, disappeared from the public eye. The only contacts between the parties were imposed by legal procedure.
- At the same time, the project's promoters were doing everything in their power to convince the opponents not to take any further legal action.

Not only did the CABR deal with the legal aspects of the opponents' action, but it also had the project entirely reassessed. Two technical experts were asked to answer a series of precise questions in the presence of the lawyers of both parties. Following this reassessment, additional requirements were requested concerning the drainage of waste water.

On the opponents' side, the lawyers R. Zweidler and H.P. Schmidt were in charge of leading the action. Mr. and Mrs Foege, also lawyers and leaders of the 'Morgenluft' group, were edged out of the battle.

Two months before the appeals commission was to meet over the issue, the deputy Gallacchi, whose ties with CG were known to all, made a proposal before the cantonal parliament. Judging by the support he received, opposition to the SWIP project remained marginal in Basel. At the same time, Gallacchi's motion made things difficult for German opponents, since the debate had shifted to the issue of cross-border collaboration.

began. The mayor admitted that these discussions had been effective, since he was persuaded to abandon legal action and negotiate directly with CG. The mayor's change of attitude, which followed the rejection of the appeal, can be explained by the fact that he considered having received sufficient guarantees concerning the reliability of the technical measures taken by CG, and that pursuing legal action could have a negative effect on the relations between the cities of Basel and Weil am Rhein, which were fragile, and when broken, difficult to mend.

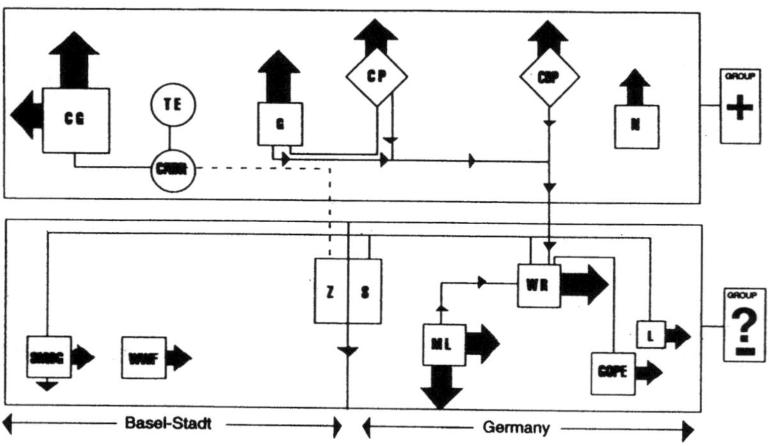

FIGURE 5.7 Network of actors. Stage 3.1. Processing the appeals, 1990 (for a key to the abbreviations see the list at the end of the document)

Stage 3.2 Direct negotiations are preferred to legal action.

The network of actors was suddenly greatly simplified (see Fig. 8).

CG accepted the mayor of Weil am Rhein's proposal, without setting any conditions. Confident in its own level of technical competence, CG also promised to lower the pollution level to well beneath both the Swiss and German legal limits and agreed to submit the completed facility to an inspection; the opponents would be allowed to participate in this procedure.

However, in its written declarations, CG was careful to ensure that its commitments could not be interpreted as a recognition of standards stricter than those enforced by the Swiss legislation.

When Weil am Rhein gave up its legal action against the project, the opponents lost their main source of funding for their lawyers. Continuing the struggle became difficult.

In Basel, direct negotiations soon took place between the opponents, the cantonal authorities of Basel-Stadt and CG. The WWF replaced the SMOG group as representative speaker of the opposition. After a few weeks of discussions, CG produced a second declaration, whereby the company accepted the presence of SMOG on the future advisory board, guaranteed the right to information concerning the results of the measures taken by the company and granted the WWF the right to evaluate the technical basis of the incineration and smoke washing facilities.

As a result, the appeals were withdrawn, including those filed by the Foege group ('Morgenluft'). The two Foege lawyers considered their withdrawal a defeat. The lawyer Zweigler also regretted that the conflict had not been brought to Federal Court, since this would have enabled the environmentalists to set a few legal precedents.

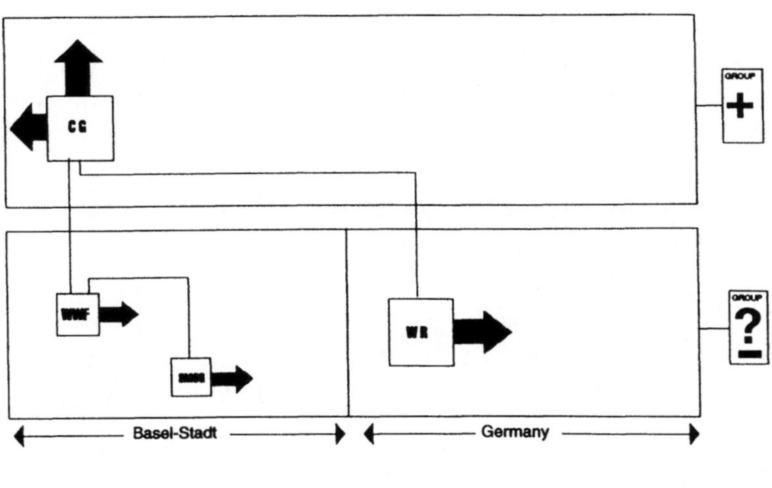

FIGURE 5.8 Network of actors. Stage 3.2. Direct negotiations after the withdrawal of the appeals, 1992 (for a key to the abbreviations see the list at the end of the document)

5.4 CONCLUSIONS: SUCCESS FACTORS

After studying the case, five success factors come to the fore. The following section shows how the roles were defined in the case of the SWIP of CG: one group adopted a dominant position, while the others adopted an attitude of resistance. How could the distribution of these roles among the actors contribute to the success of the decision-making process? An impression is given below.

1. Disparities between group + and group ?[29] in terms of stability and clarity in their action

The 'Group +'[30]

The actors of the 'group +' were quick to establish their respective roles. The relationships between the various actors were clear-cut; they were strengthened during the process and were modified only when major events occurred (such as the Schweizerhalle fire and the city of Weil am Rhein's decision to abandon legal action). All changes were justified by the necessity of maintaining or improving efficiency.
CG thus remained the leader of the project throughout the entire process, even when the issue had become political. CG's leadership was underscored by the fact that Sandoz and Hoffmann-La Roche chose to remain in the background. The DPW/BS supported the project from the start. After the Schweizerhalle fire, its role was slightly modified, in order to emphasise its image as protector of the law. The head of the DPW/BS, who previously openly supported the project, withdrew into the background.

All these actors shared a clear and common 'philosophy of action', based on established practice. This enabled them to pursue their action with determination and confidence.

Thus, during the entire decision-making process, CG managed to defend its position while presenting a project which was in fact practically

---

29 For a key to the abbreviations, see list at the end of the document.

30 This group includes, for the purposes of this report: the sponsors of the project (CG, Sandoz, Hoffmann- La Roche), the project's advocates (the authorities of the Basel-Land canton, Mr. Gallacchi and the majority of the Basel Grand Council deputies he managed to mobilise in favour of the project), the concerned Basel-Stadt public services and authorities (COPE, ICPE, CABR), and the FOPE.

devoid of technical details. The COPE accepted to go along with CG. The permit was thus granted on the basis of the brief which CG promised to respect.

The head of the COPE professed a philosophy of communication which he adapted to his objectives.

Though the roles played by the members of the 'group +' were stable and clear-cut, they were nevertheless in many ways unconventional. Thus, as any private company, CG's first intention was to build the SWIP at minor cost. However, after the Schweizerhalle fire, CG agreed to take into account the increasingly important requirements of environmental protection. In this way, CG showed that it wished to win back the Basel residents' trust in the local pharmaceutical industry, and that it had the means to do so. The FOPE was careful not to interfere in the relationships between the sponsor company and the Basel-Stadt authorities in charge of evaluating the project, but did not hesitate to openly support the CG project, a key element in the federal special waste elimination programme. The COPE played a double role : on one hand, it was officially in charge of enforcing the law and legal standards; on the other, it acted as a mediator between the proponents and the opponents of the project. As a representative of the public authorities, it participated in the elaboration of the final and detailed brief of the facility.

The 'Group ?'

Unlike the actors of the 'group +', those of the 'group ?' found it difficult to define precise roles and keep to them. The nature of these roles and the objectives to be attained were not always clear-cut and often modified. Though some protest groups did have a clear philosophy of action (e.g. the group SMOG, the Foege's 'Morgenluft' action group, the lawyers Zweidler and Schmidt), others were more ambivalent and less constant in their position (such as the president of the Land of Baden-Württenberg, the German municipalities which participated in the action along with Weil am Rhein, the group 'Aktion Selbstschutz').

But more importantly, the 'philosophies of action' of the main opponents to the project (SMOG, Morgenluft, Zweidler and Schmidt, Weil am Rhein authorities) were based on aims that only partially converged.

These groups never really moved away from their traditional role of protesters and never changed their pattern of behaviour, except at the very end of the process, during the 'negotiation' phase. However, due to their diversity, they did have trouble efficiently co-ordinating their actions.

2. Disparities in terms of flexibility and adaptability in concrete action.

The 'Group +'

Though they were guided by a constant and common philosophy of action, the actors of the 'group +' adapted their moves to each concrete situation, operating by stages and following the step-by-step institutional decision-making process (publication of intentions, preliminary application, application for a building permit, appeals). They worked according to the principle of 'trial and error' and in their strategy, they refused to anticipate beyond the present phase and the preparation of the following one. This made it possible for them to adapt their actions as best as possible to the way events were turning out. They showed a lot of initiative and often surprised their adversaries by abruptly changing tactics.

Concretely speaking, the publication of intentions came in the wake of the successful incineration of the waste from Seveso, which was considered, to a certain extent, a triumph. After the Schweizerhalle fire and the adoption of a federal law on the protection of the environment, the concrete implications of which were not yet quite clear, the preliminary application phase was used as a means to test the situation. The application for a building permit served to strengthen the project through communication and exchange of information, since the project was adjusted according to the demands of the concerned actors, usually through COPE mediation. The appeals phase was dominated by the wish of the 'group +' to be done with the process as soon as possible. As a result, the project promoters' strategy was to refuse all dialogue with the opponents, to try to cause conflict among them, and to persuade them that they would gain more by negotiating than by attempting legal action.

The 'Group ?'

The opponents, on the other hand, were not able to adapt their tactics to the changing issues of each phase of the debate. At each stage of the process, they continued to proceed and speak as they had during the preceding stages. They were thus easily taken by surprise by the new elements in the strategy of the 'group +' and ended up being forced to adapt to the circumstances and give in on a number of points.

It is true that this type of action, which is typical of minority groups, is not entirely ineffective. By sticking adamantly to their positions at first and accepting certain concessions only afterwards, the 'group ?' was able to obtain positive results despite its intrinsic weaknesses.

However, these results were made possible by the fact that the COPE shared several of its objectives with the project's opponents. Furthermore, given these circumstances, it is clear that had the opponents been able to co-ordinate their action, they would have been more effective.

3. Choosing the right basis for negotiation.

The 'Group +'

Whenever the actors of the 'group +' negotiated among themselves and with their opponents, they were always careful to chose topics where they were sure to have a solid background. They were thus able to negotiate in the full sense of the term, since they remained in control of the process they themselves had initiated and for which they were responsible. Thus, concerning technological questions, CG was ready to meet the challenge, confident that its competence on the question was superior to anyone else's in the field. The COPE was in a strong position to discuss the legal interpretation of the LPE, since as long as the enforcement ordinances were still pending, the COPE still had more power than others in this area. If this process had taken place today, now that the enforcement ordinances have been implemented and that the interpretation of the law has been codified, the COPE may not have had as much leeway as it did between 1988 and 1990 for the Basel SWIP project.

The 'Group ?'

The opponents, for the most part, also chose to fight on technical and legal grounds; however, as they later found out, the credibility of their initial position, which consisted in opposing the project on principle, was lost as a result.

The 'Morgenluft' action group adopted a very radical position and campaigned among the residents. Their aim was not to pressure CG but the authorities of Weil am Rhein. Indeed, all the latter ever asked for were guarantees that the foreseeable impacts on the environment would meet German legal standards.

4. A combination of positive factors

It is important to understand that the factors described in points 1 and 3 and said to have contributed to the success of the project were only effective because they were combined.

Indeed, clear-cut roles and concrete methods of action are not enough to ensure success if the corresponding strategies are rigid and unable to adapt to changing circumstances. Similarly, if roles are not clearly defined and if the action is not guided by a clear philosophy, an adaptable, step-by-step strategy will serve no purpose. In the latter case, the group + would have found itself unable to control the process, even if it had been in a stronger position. In both cases, the group + would have had trouble negotiating, even if it had chosen to discuss points where it had an obvious advantage.

5. The importance of context

However, a clever combination of the attitudes described above is still not quite enough to ensure success. In addition, the circumstances must be favourable to the project.

This was indeed the case for the Basel SWIP project :

- The final aim of the project was never really criticised by anyone, except in the weeks following the Schweizerhalle fire. The population of Basel was aware of the fact that a solution was needed to deal with the elimination of special waste by incineration, and that the best solution was to have a facility located near the production source. No one denied CG's competence in this field. Technologically speaking, CG had no rivals. What people feared was that in order to cut costs, CG would not use all its know-how. After the Schweizerhalle fire, however, CG had to recognise that economic considerations were of secondary importance. The accident thus created a context which made it more likely for CG to come up with an acceptable project.
- The local population, especially in Basel, is historically extremely dependent, from a socio-economic point of view, on the pharmaceutical industry. For this reason, the majority of the citizens concerned were necessarily favourable to the project. Events have shown that the Schweizerhalle accident did not have a deep impact on this attitude.
- The area's geopolitical situation was also an objective factor in the project's success : the local opponents to the project obtained much less political support than they would have had in other regions of Switzerland. The canton of Basel-Stadt represents hardly more than one municipality. Local opponents thus found it difficult to garner political support for their struggle, since they could not create a conflict between local municipal authorities and the central cantonal authorities. On the other side of the border, the opponents to the project managed to organise the movement on the municipal level, but the official action of the municipality was hampered by the border between the two countries. The German municipal authorities had no power over the Basel cantonal authorities in

charge of approving the project. As far as media attention was concerned, they were dependent on the Basel press's willingness to lend an ear to their protests, since few people in Basel read German newspapers. During the appeals phase, the Basel politicians who supported the project were able to take advantage of the 'border effect' by threatening the authorities of Weil am Rhein to cease all cross-border collaboration if they didn't withdraw.

Thus, given the nature of the objective circumstances which made it possible for CG's SWIP project to succeed, the analysis of the success factors can only be valid for projects which also boast rather favourable objective circumstances. This does not mean that the success factors described are ineffective if the circumstances are not favourable, but simply that no lessons or rules for future cases of this type can be drawn from this case study.

Two general remarks should be made to conclude this study.

The lessons to be learned from this case can be applied to other projects besides those which have an impact on the environment. They may be applied to any project which arouses political interest and concerns a field where standards and legislations abound. The Basel case shows that one of the best ways to succeed in such an undertaking is to encourage co-operation between the private sponsors of the project and the public authorities in charge of enforcing the law. The aim of such a collaboration is to establish, beyond the legal framework, specific partnership agreements. However, the question remains of whether such a system wouldn't result in a weakening of democratic rights.

Managing this type of case requires special competence which is acquired through practice and the analysis of precedents. For this reason, it would be a good idea to create a specific professional qualification enabling a person to master the technique of dealing with such cases, in the same way as an engineer is qualified to design the technical aspects of a project. A specialist in the management of conflictual decision making processes could be a winning card in the hands of the group which employs him. In any case, the question deserves to be discussed.

NOTE CONCERNING THE 'NETWORKS OF ACTORS'

- The proponents are in the box

  The opponents, or those who delay the process, are in the box

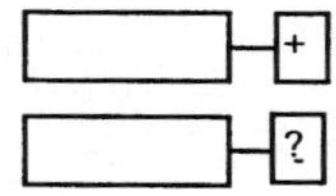

- The following figures represent the following categories of actors :
  - The actors who actievly fight for or against the project
  - The actors who verify the legal aspects (or mediate)
  - The actors, organizations, events which play the role of "arenas" or "filters" in the formulation of opinions

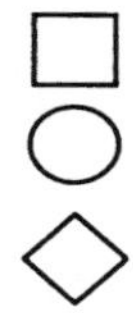

- The size of the symbols  shows the importance of the actors. The position of each group in his box in relationship to the other groups - whether it is in the forefront or in the background - reflects the nature of the role (main or secondary) played by him at a given stage.

- The arrows indicate the aims of the actors, according to their direction :

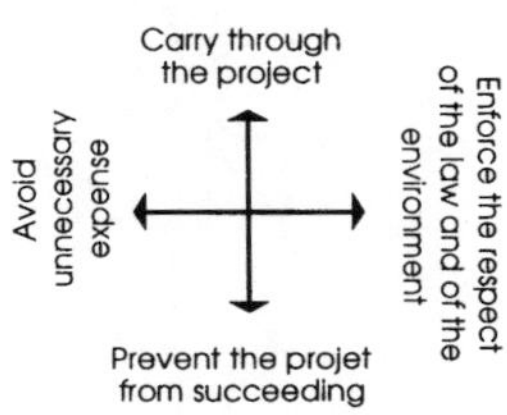

The length and the thickness of the arrows show the importance ascribed to a given aim

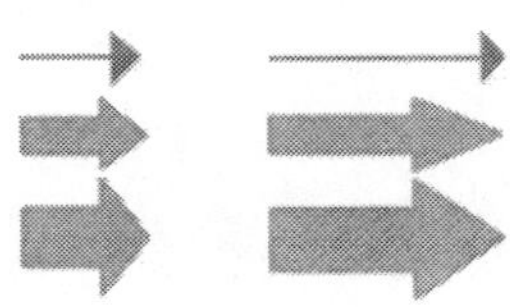

- The lines linking tho actors express the following ideas :
  - collaboration
  - communication
  - pressure

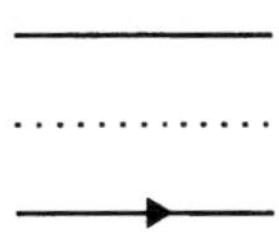

6 BALANCING THE STAKE. THE CREATION OF THE CENTRE FOR INDUSTRIAL WASTE TREATMENT IN MODENA (ITALY)

ANNA LUISE, PAOLO FARERI

6.1 THE PROBLEM

The Centre for Industrial Waste Treatment is the biggest plant of its kind in Italy, and the only one under public management. It is located in Modena, in the Emilia Romagna region, and is managed by a public agency - AMIU (Azienda Municipale di Igiene Urbana) - owned by the local municipality. As it is today, the Centre is the result of the realisation of a first plant and the subsequent addition of two other plants. Far from being designed as a single project implemented in different phases, the Centre was added on to during the years, each plant having been decided upon separately. What we have, then, is not a single process decision making, related to the Centre as a whole, but three different processes concerning:

a) the realisation of the first inertisation plant (called Soliroc 1) for toxic waste treatment, with a capacity of 25,000 tons/year (1978-1985);
b) the realisation of an incinerator for toxic waste, with a capacity of 5,000 mc./year (1984-1993);
c) the realisation of the second inertisation plant (called Soliroc 2) for toxic waste treatment, with a capacity of 50,000 tons/year (1988-1993).

A relevant consequence, from the point of view of the research, is that, as we will see more clearly in the analysis, the problem being attacked by the involved actors is not stable. Each process is related to a different problem. In defining the limits of our field of analysis we have included the three processes. This is because we think that the relevance of the Modena case is to be found in the capacity to create a centre for toxic waste treatment that is exceptional for its dimension, at least in the Italian context.

B. Dente et al. (eds.), The Waste and the Backyard, 161–174.

The initial problem is clearly local, and is determined by the concentration of industries - in the district of Sassuolo - producing tiles and ceramics, and by-producing large quantities of waste.
The perception of the problem of industrial waste dates back to the end of the 70s. In 1978 the Modena City Council approved a public dumping project for solid urban waste and industrial waste, as a result of an agreement between the most relevant single producer of industrial waste in the province (Fiat Trattori) and the municipality. Fiat Trattori financed the realisation of the plant - with a capacity of 6,000 mc., of which 4,000 mc were reserved for Fiat (on land owned by the Municipality). At the same time the USL (Local Health Unit) of Modena carried out a survey on industrial waste in the region for the first time, producing a relevant source of information that was used as a basis for setting up local policies.

The creation of the Centre - i.e. of the first inertisation plant (Soliroc 1) - is the solution to a serious local problem of industrial waste production. As we will see in the analysis, the extension of the Centre - with the realisation of the incinerator and Soliroc 2 - is less clearly a solution to a local problem, and more evidently the attempt to implement a development strategy by the local agency responsible for the management of the Centre itself. From the point of view of the AMIU, it is a partial success. From our point of view, it is a success story, and we will try to identify the relevant factors for this positive outcome in the final chapter.

The story of the Centres' creation, however, is linked to another problem of waste treatment whose source is clearly not local. In 1988 an illegal toxic waste landfill was discovered in Nigeria near the port of Koko. The extremely dangerous material was coming from Italian industries. The conflict between the Nigerian and Italian governments ultimately resulted in the decision to send the waste back to Italy for final treatment. The problem - deciding where to unload the ships returning the waste to Italy; where how to analyse the waste; and finally, where to treat it - had a very relevant impact on the political debate at the national level and generated serious conflicts along different dimensions (between political parties, the centre, and the periphery, etc.). In the end, the Modena centre was one of the recipients of waste for final treatment. The «Karin B story» (from the name of one of the ships used to take the waste back to Italy) had a relevant impact on the decision-making process for the extension to the centre, and has been considered as part of the case study in the chronological reconstruction of events. At the same time, the analysis of the network of participating actors is restricted to those directly involved in the creation of the centre.

6.2 THE CHRONOLOGY

First Phase: A Local Problem of Waste Treatment (1980-1983)
The starting point in the creation of the Centre for Industrial Waste Treatment, following the survey by the Modena USL on industrial waste production in the province, is to be found in the decision taken by the Modena City Council in November 1980, concerning the realisation of a platform for sludge treatment based on the adoption of a British polymerisation technology (the Sealosafe technology by the British company Stablex). The decision made by the Modena Municipality is partly the result of the approval by the Regional Council of Emilia Romagna in May 1990 of the Regional Law n.37, concerning «Interventions for implementing solid waste and sludge disposal plants». The law granted financial support to single Municipalities and/or associated bodies, including joint consortiums. This was the basis upon which the Modena municipality asked for financial support to set up the centre.

Following a visit by a delegation of experts (including representatives of the AMIU, USL, the Modena Province, and Municipality) to a French treatment plant where different technology - Soliroc, by the Belgian company Chemstobel - was used, a comparative study of the two technological solutions was issued by the Modena USL in 1981. After further in-depth technical studies and economical assessments, a final decision was made by the City Council on November 1981 the Soliroc technology to choose.

Just a few months before, in July, the Region had approved a financial contribution of 2 billion Lira, covering 50% of the total cost for the realisation of the plant. Works started in 1982, and in June 1983 the management of the whole programme was entrusted to AMIU.

In September, the Region approved a new contribution of about 2.5 billion Lira, covering, in the end, about 100% of the final cost. This new contribution was increased slightly in the final calculations.

In April 1984, testing of the new plant started, and the official opening was approximately one year later in March of 1985.

Second Phase: Waste as a Development Opportunity (1984-1988)

The decision to add to the inertisation plant (an incinerator) was taken by the Modena City Council just after completion of the first phase while testing was in process in October 1984. Once again, the Emilia Romagna Region financially supported the project with a contribution of 1,55 billion lira. This covered about 50% of the costs of the project.

The procedure was completed very rapidly: just after five months, in March of 1985, the project was approved and the construction began. It took ten years (beginning of 1994, however before the incinerator began to operate.

Different reasons contributed to the delay in the schedule. In a first phase (until 1989), problems were basically of a financial nature. After a stop between April and October 1985, works were suspended for more than two years: from March 1986 to May 1988. The Region had not completed the procedure that was needed to make the approved funding available.

In the meantime some changes and additions to the original project had been decided upon.

The increase in the costs for the realisation of the incinerator, and the problems in the implementation of the project did not deter those involved - AMIU and the Municipality above all - from proposing a new project: a new inertisation plant (Soliroc 2). The City Council approved the project in October of 1988. It is at this point in the development of a local waste management policy that a completely external and unexpected event modified the stake, as perceived by most of the participating actors, and consequently the decision making process itself.

What is known in the national debate as the «Karin B event» began in the summer 1988, that is, an illegal landfill - containing hazardous waste coming from Italian industries - was discovered in Nigeria. The international dispute was resolved with the decision by the Italian government to take responsibility for the treatment of the waste and to clean up the landfill area. In July of that year, the publicly owned Italian chemical company, ENI, rented a German Cargo Ship on behalf of the Italian Government. The ship left the port of Koko on July 30th taking on board the first part of the toxic waste. For about a month Karin B tried, without success, to unload its cargo.

The proposed destination was Ravenna, but the local authorities withdrew permission, worried by the negative effect that the arrival of waste could have on tourism (the main economic activity in the area). And a negative answer was given by a number of other ports, in Italy and elsewhere: La Spezia, Genova, Venezia, Cadiz (in Spain), Neath (in Wales), Plymouth, Rotterdam, some ports on the French coast, etc. At low speed, the ship started to travel back to Italy, not knowing where to head for.

The break in the impasse was determined by the initiative of the president of the Tuscany Region, who proposed the port of Livorno as the possible destination. The ship would be unloaded in a special area of the port, where the barrels would be analysed and then directed to plants for treatment. The proposal was immediately accepted by the Secretary of the Environment, and at the same time by the President of the Emilia Romagna Region, indicating the treatment plants in the region as possible destinations for the Karin B waste.

A final agreement among the Ministry and the two regions was reached in September, and was based on the identification of a number of projects areas with serious environmental problems. These projects were to be financed by the State as compensation. Modena was chosen as one of the destinations for waste treatment.
Karin B arrived in Livorno on November 29, 1988.

Third Phase: Unexpected Events, Local Conflicts and the Redefinition of Waste Treatment Policies (1989-1994)

The Karin B event, and the arrival of waste in Modena, meant that citizens began to take note of the AMIU plants, and, in particular, of AMIU projects: the incinerator under construction, and Soliroc 2, the inertisation plant extension that was recently approved.

In January of 1989, the first barrels of waste from the Karin B arrived at Modena. A strange odour began to come from the AMIU soon after the North Modena Citizen Committee was created, whose first act was to request a stop on all plans to extend the AMIU plants.

Just a few weeks before the Crocetta District had set up a local Committee on "Waste treatment and environmental protection" with the objective of facilitating the relation, between citizens and the public administration, and of distributing information about the Karin B waste treatment operation and the activities of the Centre.

After the Karin B, another event contributed to increase the level of conflict on the development of the Centre for Industrial Waste Treatment. In April 1990 the AMIU admitted that an agreement with the Bolzano Municipality had been made, concerning the treatment in the Modena plants of 260 tons of toxic waste, some of which had already been treated, before the contract had been rescinded.

It is more and more clear that the development strategy of the Centre was based on the idea of importing waste from outside for treatment. The non-local character of the problem is demonstrated - on the other hand - by the progressive reduction of waste produced by local industries: from 29,000 tons/year in 1990 to 20,000 tons/year in 1992.

In the period 1989-1991, the level of conflict on the waste treatment issue remained very high. The North Modena Committee directed its opposition on three main elements: the creation of a new incinerator for domestic waste; the creation of the incinerator for industrial waste; and the treatment of waste coming from the Karin B. In 1990, in correspondence with the local elections, the issue was at the centre of political debate.

In March of 1990 the incinerator was completed, but just after a few months, in July, the Municipality decided to realise some new works; in particular, the total closure of some containers where waste is kept before treatment. This was actually a direct answer to the requests made by the North Modena Committee, who protested because of the odour coming from the plant. Works on this part of the project would be completed in May of 1993. In the meantime, however, testing on the plant can proceed.

The Karin B event was officially concluded in March 1992. The Modena centre received 928 tons of waste. Analysis in order to identify the best treatment had been conducted during 1989. The waste was treated between July of 1990 and December of 1991.

Yet just when the controversial episode had finally come to a happy end, the local opposition succeed in its struggle against the incinerator. On May 14, 1992 the Emilia Romagna Administrative Court accepted the lawsuit by 46 residents in the Crocetta neighbourhood, and there by annulled the authorisation given by the Modena Province to AMIU to start operating the plant for the testing.

The lawsuit was based on differences in the interpretation of the existing legislation concerning the emissions limits in the atmosphere by waste treatment plants, where different limits are given to new plants or extensions of existing plants. The Modena Municipality applied to the State Court in order to fight the decision of the Emilia Romagna Administrative Court. Meanwhile, the extension of the inertisation plant (Soliroc 2) was completed in June of 1993.

In the same year, the controversy surrounding the incinerator came to a close , though not without some residual problems: the Modena Province issued a new authorisation to start operating the incinerator. This led to the testing phase. But the first tests during 1993 failed because the mix of waste used in the testing was different from what was presumed in the design of the plant. The plant was supposed to treat imported waste of different types. The second set of tests in 1994, was positive and resulted in a normal operation of the new plant.

6.3 Actors, Networks and the Patterns of Interaction

First Phase: An Efficient Team Work (1980-1983)

The first phase of the decision making process was (as shown in the chronology) very simple. It seldom occurs that a decision-making process corresponding so precisely to the formal procedure. All the events identified in the analysis pertain to the procedure itself; all the participating actors are the ones with some sort of formal responsibility in the design, decision, and implementation of the project; and all the actors are apparently play the role they are expected to play. Finally, all the actors with an assigned role in the procedure are actually mobilised in the course of the process, and join positively the network.

During this first phase, the main axis of interaction is defined, the one that will structure the decision making process as a whole: a line with the Region and AMIU at both ends, and the Modena Municipality in the middle. Other actors join the network in the different phases around this main axis. During the first phase, the list of participants includes no more than the private company realising the plant, the other institutional bodies responsible for relevant steps in the procedure and the experts involved in the design of the technical, organisational and legal aspects of the project.

The pattern of interaction is quite clearly problem solving: the participants share the same definition of the problem (a local problem of waste treatment) and co-operate in order to identify the best available solution, through a clearly defined division of roles and duties.

Success in this phase - that is, in the realisation of the first plant - is largely determined by the increase of the stake (or more probably, the creation of a 'real' stake) due to the intervention of an actor from outside the local network: the Region with its legislative act and, of course, with its economic resources. The positive sum nature of the local game in the absence of local opposition to the plant, is then guaranteed: the municipality can improve its intervention on environmental protection at no cost; AMIU can increase its field of activity.

Within this framework, the public administration works is at its highest standards, though this, of course, is not the rule. The Modena municipality is qualified enough to mobilise the necessary expertise in order to guarantee the identification of the right solution (in technological, economical, and legal terms), be it within the public bodies involved in the procedure, or in the hands of external experts. The municipality is also efficient enough to implement the decision - in a completely new field of activity at the local level - in a more than reasonable time.

Second Phase: Diverging Objectives in the Public Sphere (1984-1988)

The second phase of the decision making process is more like an interlude between the first one (when a local hazardous waste treatment policy is framed, and the first plant is realised), and the third one (when conflict arises, and the most relevant outcomes are defined). Relevant things happen during this phase but the consequences will become visible only in the subsequent years.

It is during the second phase that the development strategy of the Centre for industrial waste treatment is designed, dash a strategy based on a definition of the problem which is different from the past, and, as we will see, not shared by the main actors involved.

The realisation of the first plant changes the nature of the game at the local level: mobilised actors remain the same, but with different objectives, and consequently, with different patterns of interactions.

Although not clearly shown in the analysis, the relationship between the Municipality as the promoter/director of the process and the AMIU as the executor seems to be subverted. Waste is not a local problem to solve from the point of view of the local agency, but a business opportunity to develop. The role of promoter in the network is more likely played - in this phase - by AMIU, while the Municipality seems to be the place where a political consensus on those objectives is constructed. This change of roles is facilitated - once again, during this phase - by the permanence of the Region as the actor injecting resources from outside the local arena, giving the game a positive sum shape.

The idea of a 'centre' for hazardous waste treatment implies the presence of different plants adopting different technologies to treat different types of waste. This is what happens with the project of the incinerator after the first inertisation plant (Soliroc 1) is completed. On the other hand, it is the success of Modena as a waste treatment centre of national significance (largely a result of the «Karin B event») that reinforces the idea of a further development of the centre itself.

Strategies and events contribute to deepen the distance between local problems and the role of the Centre, dash a tendency that is reinforced by the changing trend in waste production in the tiles industrial district of the Modena region. It is in these years that local industries are defining and adopting alternative technologies to reduce the amount of waste produced, and thereby increase recycling.

While the Centre's treatment capacity is dramatically increased by the new plants, and while Modena is seen as one of the few existing solutions to national and exceptional problems of waste treatment, the local demand is prospectively decreasing.

This is the situation when new actors enter the local arena, opening the third and last phase of the process. Less relevant, for what concerns the overall interpretation of the decision making process, is the length in the implementation of the incinerator project, due - as it seems to be - to factors independent of the behaviour of the participating actors.

Third Phase: The Construction of a Set of Rules (1989-1994)

The third phase of the decision making process is characterised by the appearance of new actors (citizens groups) on the stage, by the rise of conflict, and by the strategies enacted by the central actor in the network

- the Modena Municipality - in order to solve the conflict and, so, to avoid the risk of paralysis.

The final outcome is that the new plants are completed and operating, while at the same time a new set of rules is defined, concerning the use of the plants and the monitoring. Both the plants and the rules represent the success of the decision-making process.

The North Modena Citizen Committee objectives are very clear. Created as a consequence of the decision to treat in Modena part of the hazardous waste coming from the Karin B, the local group tries to stop the development of the Centre for Industrial Waste Treatment, and in particular the completion of the incinerator. The Karin B event, apart from the specific issue of danger and the emotivity that characterises the story at the national level, unveils the shift in the role and scope of the Centre. No a facility for the treatment of a local problem, but, in the mind of the AMIU at least, it becomes a facility for the treatment of hazardous waste (regardless of its source).

The initiative by the Committee is very successful, both in terms of the support at the local level, and in terms of the impact on the media and the public debate. The promoters use different types of resources, and in particular - as is often the case in these stories - legal resources, in order to raise procedural and technical objections in order to block the realisation of the incinerator. They first succeed, when the administrative court accepts their appeal, but fail in the end.

The patterns of interaction, in the conflictual phase of the process, are very well defined by the actors themselves. The North Modena Committee identifies the Municipality as the exclusive counterpart, continuously reaffirming the competence of the AMIU as the technical executor of the objectives defined by the City Council. The relationship between the Committee and the Municipality is crucial.

In the managing conflict, the behaviour of the Municipality during this phase is based on three main elements. The direct relations with the local opposition group are, and remain, conflictual. Modena administrators accept the confrontation game with the North Modena Committee, labelling the group's action as 'localistic', 'fragmented', and instrumentally supported by the opposition within the City Council.

At the same time, the governing coalition objectives are directly influenced by the action of the committee. Both the Karin B and the incinerator issues are obviously relevant in the public debate when, right in the middle of conflict, the time comes for the local elections, in 1990. The new coalition[1] consistently changes the set of objectives in regard to the Centre and its development, including in the programme the most relevant elements emerged from the conflict:

a) the clear delimitation of the area served by the Centre (the province of Modena), excluding the possibility of importing waste from outside;
b) the definition of a plan for the further reduction of impacts of the plants on the nearby neighbourhoods;
c) the realisation of a monitoring system, whose results will be widely available to the public, and in particular the residents in the nearby areas.

What remains in the hands of the local opposers to the incinerator, once the City coalition has redefined its objectives, is just the more evidently 'nimbystic' stances, the most difficult to legitimise in front of the public opinion. In this way, the legal technicalities used to stop the realisation of the incinerator, not adequately supported in terms of objectives, are more easily solved by the promoter.

The third element, directly connected with the previous one, has to do with the role of promoter in the network, a role that comes back to the Municipality. The clearer message included in the programme by the new coalition is that the objectives for local waste treatment policies cannot be defined by the agency responsible for treatment - whose interests are clearly toward an increase of the treatment capacity, regardless of the nature and dimension of the local problem of waste production - but are the sole responsibility of the City Council. This of course leaves the AMIU with some non-marginal problems, for what concerns the management of the Centre.

A first problem has to do with the adopted technology for waste treatment. The test failure is determined by the fact that the plant had been designed for treating a mix of waste (including the waste produced in the area and the waste imported from outside), that was not the waste used in the tests (and, presumably, the waste to be treated in the future). A second problem has to do with the management of the Centre: the plants capacity, at the end of the story, is apparently more than the quantity of locally produced waste.

[1] where 'new' does not mean 'different' in terms of political parties relations: the Communist Party retains the majority of votes, and runs the Municipality alone.

Another interesting aspect of this phase, even if its impact on the final outcome of the decision making process is quite limited, is the reaction by the local Communist Party to the bottom up mobilisation of citizens against the waste facility project. Facing the problem of a self-organised grass-root opposition in a neighbourhood with a very high percentage of voters, the party tries to create another local committee as a sort of experiment in public participation and consensus building on environmental policies at the local level.

The Committee tries to build a common position on waste treatment policies and to distribute the available information on the plants and their expected impact on the area. Its mixed composition - members of the Committee are: citizens elected during a public meeting; experts from the AMIU, the USL, the Municipality; city councillors - is a relevant aspect of the strategy.

6.4 SUCCESS FACTORS

In the analysis of a decision making process that develops almost uneventfully, (apart from a certain degree of turbulence in the more recent years) the risk is to forget that the decision (or three separate decisions) being made is about the siting of a centre for the treatment of hazardous waste, including an incinerator. And that this decision is being made in Italy in a period during which the level of conflict in environmental policies is very high, the Nimby syndrome is spreading all over the country, and the paralysis of decision making processes on these issues is more the rule than the exception. Moreover, the story includes (and this too is a success) the treatment in these plants of part of the Karin B waste, probably the most controversial event concerning hazardous waste in Italy in recent years.

The Modena case is no doubt exceptional. It is also a success story because all the plants have been realised and are operating. It can also be considered a success story because a set of rules is eventually defined as an outcome of the process, which is recognised by the participating actors as a point of balance among the different objectives mobilised by the process itself.

Some general factors can be identified in the attempt to find the reasons for this success.

The first one has to do with the timing of the process, where the relevant aspect is the first decision occurring when the sensitivity of the public opinion on matters like waste treatment is, at least in Italy, still not very high.

The presence of the first, already operating plant has probably the effect of reducing an emotional approach by local people toward the siting of the second and third plants: there is emotion, of course, there is conflict, but it is not as much as we could expect.

The second reason has to do with the resources available to the main actor of the decision-making process; that is, the Modena Municipality. Political stability is a very important reason, not only in terms of the composition of the coalitions during the considered period (throughout the story the City Council is in the hands of the Italian Communist Party, now Democratic Party of the Left), but in terms of the capacity of the different elected officials (mayor, head of the Department of the Environment, etc.) to carry on the work started by their predecessors.

The third reason has to do with the willingness expressed by the Municipality to mobilise expertise in the course of the process from the different public bodies involved or from external experts; however, there are factors more directly related to the decision making process.

First, the presence of an actor like the Emilia Romagna Region, contributing sensibly in financial terms to the creation of the Centre. In the end, only 50% of the costs of the incinerator are paid, by local actors. The positive sum nature of the game is largely determined by the Region intervention (and the political homogeneity between Municipality and Region has probably facilitated the process).

Secondly, the capacity of the central actor, the Municipality, to manage the conflict in the crucial point of the process; that when the incinerator is under construction, the inertisation plant extension had just been approved, the Karin B waste was arriving at Modena, local conflict arose, and local elections came.

A management capacity that is based on some relevant resources (political stability, competence) but, at the same time, on the strategic interaction with the other actors involved in the process.

The main element of the strategy enacted by the Municipality is probably the mix of confrontation with the local opposition group together with the openness for what concerns the legitimate issues raised by the committee, which very rapidly become very rapidly part of the set of objectives of the Municipality itself.

Even in a situation where political consensus seems to be guaranteed (Modena is traditionally one of the 'reddest' areas in Italy, and one where elections looks more like a formality) success is clearly linked to the understanding that consensus has to be constructed. In the axis that defines the relations among the relevant actors in the process, the Municipality quite clearly gives more priority to its relationship with the community, and less priority to its relationship with the AMIU. And it is probably this choice that makes the Centre for Industrial Waste Treatment a success.

7 ANY OBJECTIONS AFTER ALL? DECISION MAKING ON THE SITING OF A DOMESTIC WASTE INCINERATOR IN WIJSTER, THE NETHERLANDS

JOSEE LIGTERINGEN, HANS BRESSERS

7.1 THE WIJSTER PROBLEM

The initiative to realise a waste incinerator facility was taken in 1986 by the waste treatment company 'Vuilafvoer Maatschappij NV' (VAM) in the village of Wijster. This village is incorporated in the municipality of Beilen, which is located in the province of Drenthe. The VAM company has a long tradition of composting the Dutch biodegradable waste fraction. The compostable fraction is taken there by train from all over the country. As a result of terminating the traditional process, the waste fraction that had to be deposited (at the same location) increased tremendously, which resulted in an increase in the deposited volume of 6 million cubic meters in about 7 years. At that time the existing landfill had nearly reached its maximum capacity. In order to reduce the fraction of the processed waste that had to be dumped, the VAM company proposed to build a waste incineration facility. The incinerator was planned to reach a maximum capacity of 350,000 tons/year of domestic waste. In order to build this incinerator several phases had to be passed: the environmental impact assessment, the provincial waste plan, the region plan, and changing the local land use plans. These procedures developed partly in parallel in a co-ordinated way.

The central problem during the decision making process is whether the waste incinerator has a positive social value, taking into account all costs and benefits. The costs of a future incinerator are especially recognised by the neighbours of the future installation. Costs are, among others, seen to be public health risks, decreasing value of houses and agricultural land in the neighbourhood, visual pollution of the landscape and increasing traffic development. The incinerator's benefits are especially recognised by the

B. Dente et al. (eds.), The Waste and the Backyard, 175–195.

VAM company, the province of Drenthe and the municipality of Beilen. Benefits of a future incinerator are reducing the waste volume, economic development of a region with a high unemployment rate, and continuation of a national industry.

7.2 CHRONOLOGY OF THE WIJSTER INCINERATOR PROJECT

In the chronology of the siting of the Wijster incinerator we distinguished three phases in the decision making process:

1. Initiative and plan formulation, which consisted of the initiative for building a waste incinerator by the VAM company, an environmental impact assess ment on the alternative options and consultation by the VAM company of province and municipality.
2. Integration of waste incineration as a purpose of provincial policy, which consisted of realisation of the provincial waste plan and the region plan.
3. Integration of a future incinerator into municipal and provincial policy against the objection of neighbours, which consisted of the realisation of municipal land use plans and the application of permits by the province against the objections by the neighbours.

First phase (Early 1986-Early 1989): the initiative for building a waste incinerator, environmental impact assessment and consultation of province and municipality.

The original initiative to realise the new waste facility was taken in 1986 by the waste treatment company 'Vuilafvoer Maatschappij NV' (VAM) in Wijster, The Netherlands. The VAM company started co-operating with the energy company IJsselmij in 1986. Besides financial reasons for the co-operation between the two companies, VAM also wanted to regain energy from the incineration process in the new waste facility. In September 1988, the VAM company sends an initiation note for an environmental impact assessment procedure to the province. In this note, the VAM company presents its plan to build an integrated waste facility, with a capacity of 350,000 tons of waste year. The note describes as an advantage of Wijster as a site the already existing infra-structural facilities and the environmental permits the company already holds. The plan also develops some alternatives for the proposed activity: the zero option, the integral burning option, the dump option and the least environmental impact option. Dutch legislation requires that these alternative options need to be discussed. In the meantime, the province of Drenthe is preparing its provincial waste plan for public inspection. In 1987 it presents an initial note for the environmental impact assessment concerning the provincial waste plan. In October 1988, the province of Drenthe decides to co-ordinate the procedures concerning the application for the environmental permits for the waste facility. After publishing the note in September 1988, the working group of the national Commission for environmental impact assessment gives its advice concerning the guidelines for the Environmental Impact Statement, part of the Dutch environmental impact assessment procedure. The advice does not invoke an order to study the possibility of alternative locations, even though such a study is a common recommendation in siting procedures.

The first phase consists of the elaboration of the ideas of the VAM company concerning the siting of a waste incinerator in Wijster. This phase starts in early 1986 and proceeds until early 1989. The problem in this phase of the decision making process is formed by realising the waste incineration by the VAM company in such a way that it can be integrated into the provincial and municipal policy. In this phase, a broad consultation takes place between the VAM company as an initiative taker on the one hand, and the province and the municipality on the other. Further, an environmental impact assessment is started in order to analyse the environmental effects of each of the alternatives to incineration.

Second phase: (Early 1989-End 1989): integration of waste incineration as a purpose of provincial policy.

In late January 1989, the province of Drenthe states its requirements for the Environmental Impact Statement. These are derived from the advice from the Commission for the environmental impact assessment. The requirements include a motivation of the choice for Wijster as the location. In January 1989, the city of Beilen agrees to the requirements. From May until August 1989, a draft version of the provincial waste plan and a draft Environmental Impact Statement concerning the provincial waste plan are under consideration by the province. Anyone can express their opinion on the plan or the statement, either writing or orally, during the comment period, and at the public hearings. On December 20th 1989, the province finally submits the provincial waste plan. As a result of the comments received during the citizens' participation procedure, the advice of the Commission for the environmental impact assessment concerning the Environmental Impact Statement-provincial waste plan and as a result of the assembly of the province council commission on environmental policy, the draft version of the provincial waste plan is modified. A major modification is the incinerating capacity of the future installation. The proposed capacity in the start note was 350,000 tons/year. The modified plan specifies a maximum capacity of 400,000 tons a year. Because of the aggravated national norms for smoke gas cleaning the provincial waste plan is modified on this point. Meanwhile the city council of Beilen finally agreed to the amendment to the land use plan 'VAM-site and surroundings' on January 26th 1989. The plan contains an exemption to the allowed height of the dump ground from 25 meters to 40 meters, which is necessary because of the increased amount of processed waste. The plan is sent to the province accompanied by a landscape-fit plan, one of the requirements of the province. The province agrees to the land use plan 'VAM-site and surroundings' in July 1989. Because the plan does not explicitly state the precise location of the proposed greenery in relation to the landscape-fit plan, the farmers' association and some neighbours of the proposed incinerator appeal in September 1989 to the Crown against this decision of the province.

As a result of the co-ordination agreement, the land use plan 'VAM-GAVI' is submitted for public inspection at the same time as the Environmental Impact Statement GAVI-Wijster and the applications for the necessary permits. Late January 1990, the Environmental Impact Statement GAVI-Wijster and the applications for the environmental permits prescribed by the Waste Act, the Groundwater Act and the Pollution of Surface Waters Act are sent to the province by the VAM company. In its initial comment, late in February 1990, the Commission for the environmental impact assessment describes the Environmental Impact Statement GAVI-Wijster as a readable document that contains a lot of relevant information, but which lacks information on specific issues that are crucial to an adequate decision making process. As a reaction to this initial comment of the Commission for the environmental impact assessment, the province declares the Environmental Impact Statement GAVI-Wijster unacceptable in March 9th 1990, and as a result of this, the applications for the environmental permits were dismissed. Only a few weeks after this the VAM company sends a report containing additional information. May 1990, the VAM company sends a second modified version of the Environmental Impact Statement GAVI-Wijster to the province. The Commission for the environmental impact assessment considers this information as additional. Meanwhile, the two applicants VAM and IJsselmij agree on the financial aspect of the waste incinerator. They both agree to pay half of the expected costs. The costs are estimated at 750 million Dutch guilders at that time.

In the second phase of the decision making process the province draws up its policy objectives. The province integrates waste incineration as a policy objective in its provincial waste plan. In its region plan the province recognises the VAM location as a possible site for the purpose of waste incineration. This phase starts at the beginning of 1989 and ends at the end of 1989. The problem in this phase consists of the integration of the VAM company's plans into the provincial policy basics. The VAM company's plans are taken as a starting point. The main actor in this phase is the province of Drenthe. The province itself has to integrate the plans into its waste and region plan.

Third phase (Spring 1990-Spring 1993): integration of a future incinerator into municipal and provincial policy under objection of neighbours.

On May 2nd 1990, the province starts three participation procedures at the same time: the permits procedures, the environmental impact assessment procedure, and the procedure concerning the draft version of the land use plan 'VAM-GAVI'. At the same time objecting farmers unite in the 'Agricultural Working Group Waste Incinerator VAM'. The purpose of this group, is to stop the siting of the proposed incinerator and if that is not possible, to seek compensation payments. A second protest group is founded in this period. It calls itself protest group 'Against Waste Incineration Drenthe', and it consists of concerned neighbours. The VAM company organises some information meetings on their plans to build the waste incinerator in this period. The participation procedures result in more than 2300 objections to the application for a Waste Act permit. The procedure also results in nearly 3000 objections to the draft version of the land use plan 'VAM-GAVI'. On May 21st 1990, the provincial council sub-committee on environmental affairs meets to discuss the Environmental Impact Statement GAVI-Wijster and the permit applications. During this meeting, the former director of the Biological Station in Wijster (who is closely related to the 'Agricultural Working Group Waste Incinerator VAM') objects to the expected calculation of the accumulation of dioxins in the neighbourhood of the future incinerator that has been predicted by the Environmental Impact Statement. The VAM company takes the initiative to test the model for calculating concentrations at an independent research institution. As a result, the statements of the former director of the Biological Station are not proved to be correct.

Towards the end of June 1990 the province finally approves the region plan. The plan contains some relevant new requirements. It mentions the VAM location as the only possible location in the whole region for the treatment of domestic waste. A second requirement is that incineration and composting of waste is under discussion and that the VAM location is considered a possible site for an incinerator. At the request of the Commission for the environmental impact assessment the VAM company publishes its final modified version of the Environmental Impact Statement in July 1990. The province and the Water Purification Board announce in September 1990 their intention to grant the permits requested under the Waste Act and Pollution of Surface Waters Act. More than a hundred objections are made to the draft version of the Waste Act permit. In November 1990 the provincial waste plan passes on approval to the Crown. At this time the city of Beilen approves of the land use plan 'VAM-GAVI'. For reasons of care and for policy reasons the province makes it clear that it gives no 'declaration of no objection' for the time being. In November 1990 the province of Drenthe gives its permission for groundwater extraction in order to build the incinerator by the VAM company. In December 1990 the province also gives its permission in relation to the requested Waste Act permit. The regional council of the national farmers' association appeals to the Crown against the final granting of the Waste Act permit by the province late in January 1991. The council also appeals the approval of the land use plan 'VAM-GAVI' by the city of Beilen with the province. Near the end of January 1991, the Crown decides not to approve the land use plan 'VAM site and surroundings'. The reason for its decision is that there are no guarantees on the exact location of the proposed greenery in relation to the landscape-fit plan. The landscape-fit plan is closely related to the land use plan 'VAM-site and surroundings'. Because of this, the Crown does not approve of exceeding the maximum dump-height of 25 meters. The modified version of the land use plan 'VAM site and surroundings' is approved by the city council on June 27th 1991. In the meantime, the Crown honours the requested suspension of the provincial decree on the Waste Act permit because of objections to the judgement of the environmental aspects.

In October 1991, the city of Beilen grants the VAM company a variation of the land use plan 'VAM site and surroundings'. On November 22nd 1991 the Crown finally dismisses the appeals against the Waste Act permit. On February 2nd 1993, the Crown finally approves the two land use plans. Next, a building permit is applied for by the VAM company. The protest group 'Against Waste Incineration Drenthe' objects to the proposed issue of the building permit by the municipality of Beilen. In June 1993 the Crown approves the issue of the building permit by the municipality of Beilen. The actual building starts in August 1993. The facility is proposed to be operational in the middle of 1996.

The third phase develops around two arenas that develop simultaneously but more or less separately after the second phase. On the one hand the decision making process consists of integrating the chosen location into the land use plans 'VAM site and surroundings' and 'VAM-GAVI' and of granting a permit according to the Physical Planning Act and the Waste Substances Act. On the other hand, the third phase deals with the objections of neighbours to the change of land use plans and to the granting of a permit and the procedures that result from it. Both parts of the third phase differ in the central problem that is dealt with, the central and peripheral participating actors, the type of activity of actors, and the main type of interaction. In one arena around which the third phase in the decision making process develops, the objections of neighbours to the change of land use plans and against the granting of a permit, results in a delay of the definite integration of the location in land use plans and permit grant. This phase runs from the spring of 1990 until the spring of 1993. The problem in this part of the third phase of the decision making consists of integration of the site into the municipal and provincial policy while neighbours continue to object. The other part of the third phase, as the other part, involves the period from the spring of 1990 until the spring of 1993. The problem in this part of the decision making consists of integrating the siting of the waste incinerator into municipal and provincial policy, while neighbours object and appeal to a higher court against the decisions made.

7.3 ACTORS

The main actors participating in the decision making process are the VAM company, the municipality of Beilen, the province of Drenthe, and the neighbours of the proposed site. Other actors involved in the decision making process, but playing a less important role, are the Crown, the Chamber of Commerce, the Commission for the environmental impact assessment, and the former director of the Biological Station. Here we will discuss the actors' objectives, their resources and their pattern of interaction during each phase of the decision making process.

First phase (Beginning 1986-Beginning 1989): the initiative for building a waste incinerator, environmental impact assessment and consultation of province and municipality.

The main actors in the first part of the process are the VAM company, the province of Drenthe, and the municipality of Beilen. Moreover, during this phase the Commission for the environmental impact assessment is involved in advising the province on laying down guidelines for the environmental impact assessment and later on the Environmental Impact Statement itself.

The VAM company plays an important role during this phase because of its initiative to develop plans for a waste incinerator. The VAM company's problem consists in the first instance of the continuity of its business. To solve this problem the VAM company has the objective of guaranteeing the continuity of its business by siting a waste incinerator. The main resources that influence the VAM company's behaviour are economic. The VAM company's position is mainly supported by financial resources and the economic implications of its business for the region. Furthermore, political resources play a role, consisting of the support that the VAM company gets for its position in the network from the province and the municipality. Finally, the VAM company has interactive knowledge, which forms the basis for its strategic acts.

During this first phase the province of Drenthe does not see itself as being confronted with a problem at all. Its objective is to apply the national governmental policy line in its own policy by means of creating a facility for waste incineration within the province of Drenthe. The municipality's problem consists at first of the difference between the objectives of various actors, which advocate alternatives that are not consistent with each other.

The actors in this phase all interact with each other. Their interactions are on the whole characterised by problem solving. The interaction between the VAM company on the one hand and the province and municipality on the other can be characterised as a positive-sum game. Both parties benefit from the realisation of a waste incinerator and on the whole will suffer costs if the project is not continued. The interaction between the VAM company and the province and municipality is classified as problem solving. The province is confronted with the national government's point of view, which gives priority to the incineration of waste rather than landfilling. Moreover, it is confronted with the VAM company's proposal to erect a waste incinerator. The province, which can only diverge from this national priority if it has a good reason to do so, has no conflicting objective with respect to the VAM company's objective. Integration of these objectives is possible. The situation with the municipality is nearly the same. The province has integrated the national priority of incineration into its provincial waste plan. The municipality can only diverge from this priority in case it has important environmental reasons. The objectives are not in conflict here, either.

During this phase of the decision making process the VAM company is the central actor in the network. The interactions that take place are concentrated around the VAM company. In this phase of the process the neighbours of the proposed incinerator are not involved as an actor in the network. Schematically the early network can be described as follows:

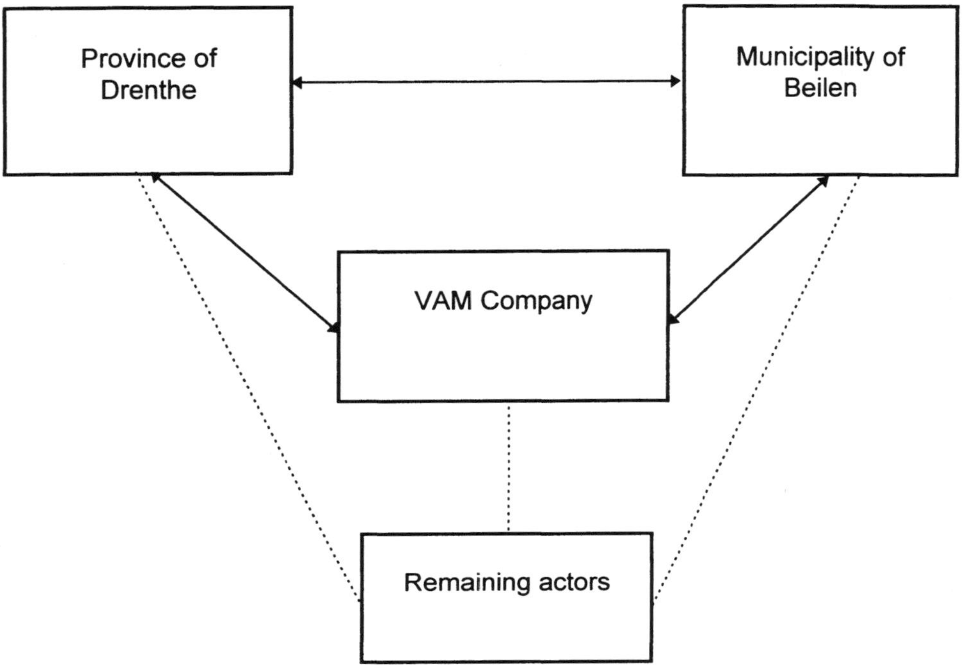

FIGURE 7.1: Interaction pattern in the first phase of the decision making process

Second phase: (Beginning 1989-End 1989): integration of waste incineration as a purpose of provincial policy.

During the second phase of the process we can recognise three important actors: the province of Drenthe, the municipality of Beilen and the VAM company. The main actor in the second phase is the province of Drenthe. The province itself has to integrate the plans of the VAM for building a waste incinerator into its waste and region plan. This can be seen as the objective of the province during this second phase. The most important resources of the province of Drenthe have a political and legal character. The province gets support from the VAM company and the municipality. Its legal resources consist of its authority to grant the permit application.

During the second phase the municipality has two options for actions: supporting the site or not. The chance of success of not supporting the siting process is estimated by the municipality at almost zero or very low, due to the fact that if the municipality does not agree with siting the Minister can (and probably will) interfere in the process. As a result the incinerator still can be sited. The municipality's main resources have a political and legal character, as have the province's. The municipality gets support for its position from the VAM company and the province. Its legal resources consist of its authority in the changing of the land use plans.

During this phase in the process the VAM company plays a passive role in relation to the former phase. Its plans are under the inspection of several actors. So the VAM company merely plays a role in illustrating and gaining support for their plans from other actors in the network.

In this phase of the process interaction takes place between the province on the one hand and the other main actors in the network on the other. The province forms the central actor in the total structure of the network at this phase of the process. The interaction between the municipality and the province of Drenthe is characterised by a positive-sum game, because both parties benefit by the realisation of a waste incinerator and only experience costs from not continuing the project. The interaction pattern between these two actors has a problem solving character. To reach an outcome of the decision making as soon as possible these actors co-ordinate their procedures concerning the changing of the land use plans and those concerning the permit application.

At this second phase, the neighbours of the proposed site are still not involved in the network as an actor in the decision making process. The interactions are characterised by problem solving. Schematically the network in the second phase can be described as follows:

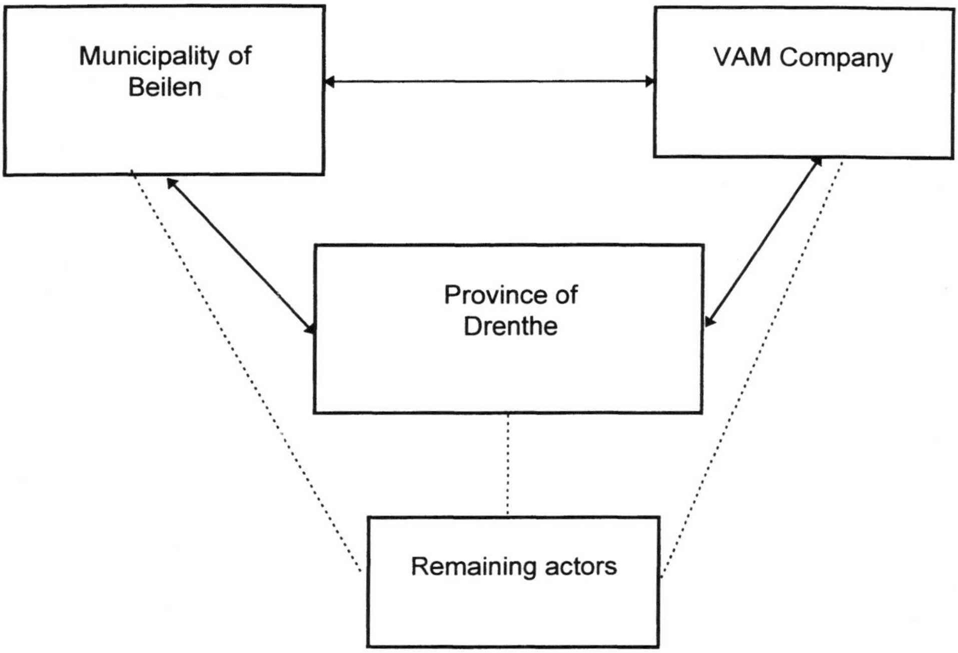

FIGURE 7.2 Interaction pattern in the second phase of the decision making process

Third phase (Spring 1990-Spring 1993): integration of a future incinerator into municipal and provincial policy under objection of neighbours.

During the third phase of the decision making process we see four main actors: the neighbours of the future incinerator, the municipality of Beilen, the province of Drenthe and finally the VAM company. Other actors that play a less important role in this phase are the Crown, who deals with the appeals in court, the Chamber of Commerce, who first seems to agree with siting and later turns to opposition, but in the end wants to get a directorship of the VAM company, and the former director of the Biological Station in Wijster. The former director of the Biological Station in Wijster is closely related to the 'Agricultural Working Group Waste Incinerator VAM'. He objects to the calculation of the expected cumulation of dioxins in the neighbourhood of the future incinerator that has been predicted by the Environmental Impact Statement. Our third phase develops around two arenas that develop simultaneously but more or less separately. Both parts differ in the central problem that is dealt with, the central and peripheral participating actors, the type of activity of actors and the main type of interaction.

The problem in the one part of the third phase of decision making consists of integration of the siting into the municipal and provincial policy, while neighbours object. During this phase the objections cause a delay to the definite realisation of the incinerator. This is why the VAM company is confronted with unexpected delay in the project's implementation. As a result the VAM company has the objective of passing through the procedures involved as soon as possible.

This is the first phase in the decision making in which the neighbours participate as an actor, besides the municipality of Beilen and the province of Drenthe. Neighbours in this part of the third phase participate in a peripheral way: they try to gain publicity for their ideas, and by doing this they more or less function as a noise producer. The municipality and the province play a central role in this part.

The municipality of Beilen is confronted with the problem that its decision does not agree with the interests of the inhabitants, as viewed by them. This problem perception results in different objectives. Within the objective of keeping its role in the process the municipality attempts to get support for siting a waste incinerator from its inhabitants. The province is confronted with a similar problem during this phase. Its decision does not agree with the inhabitants' wishes. At this phase of the process the province strives to get support for its decision from its inhabitants.

This part of the third phase is characterised by communication. The VAM company plays a less active role in this phase. Nevertheless, interactions take place between all the actors in the network. The main type of interaction in this part of the third phase is problem solving, because in this phase the VAM company's plans are integrated into municipal and provincial policy. Moreover, co-ordination takes place between the municipality and the province concerning the procedures of changing the land use plan and of the granting of permits. Schematically the network in this part of the third phase can be described as follows:

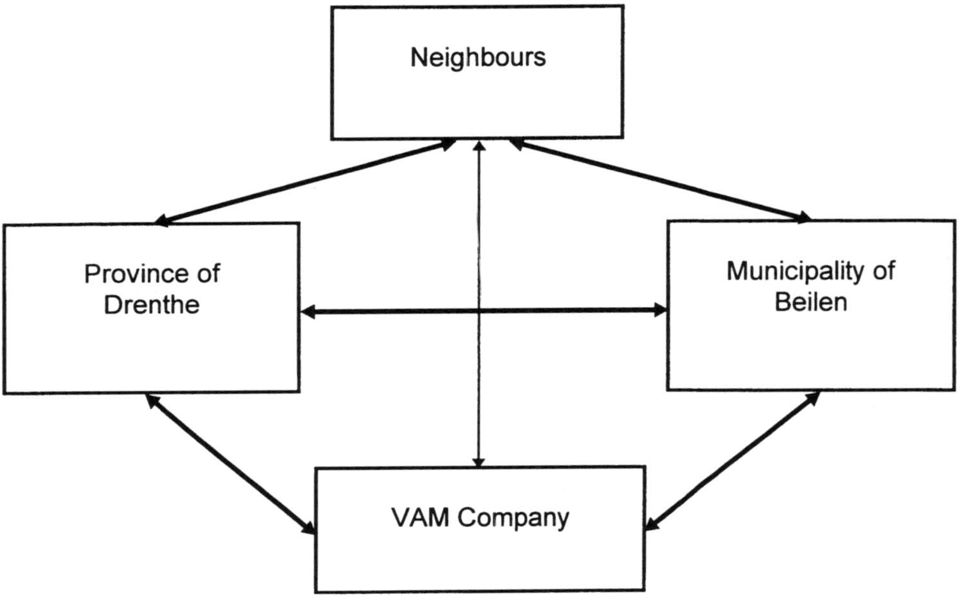

FIGURE 7.3A Interaction pattern in the plan stating part of the third phase of the decision making process

The problem in the other part of decision making process consists of integrating the siting of the waste incinerator into municipal and provincial policy, while neighbours object and appeal to a higher court against the initial decisions. Furthermore, the difference between this part and the other part is the fact that the neighbours and their organisations are the main actor in this part of the third phase. They take the initiative to appeal to the court against the decisions made. In this way there is also a difference between the type of activity that is taken in the both parts of the third phase.

When they enter the process the neighbours of the proposed site are confronted with the problem that they expect public health effects from the proposed waste incinerator. As a consequence they have the objective of stopping the siting process. Later the problem perception of the group 'Against Waste Incineration Drenthe' changes to the view that waste incineration is not a sustainable way to reduce the volume of waste. This does not result in a change in their objective. The vision of 'Agricultural Working Group Waste Incinerator VAM' shifts to considering the problem more as a cost problem resulting from the siting of a waste incinerator. As a result the Working Group strives to agree compensation arrangements with the VAM company.

The legal resources are the most important ones that the neighbours of the proposed incinerator have. Dutch legislation gives them the right to participate in consultation procedures and to object to and appeal to the court in the procedures on permit grants and changing land use plans.

The municipality and the province also play an important role in this part as defendants of the decisions made. This results in not taking the initiative of direct interaction with the neighbours. In contrast to the other part, this part is not characterised by communication. In this way the type of activity of specific actors in this part differs from the type of activity in the other part of the third phase. The VAM company, as it does in the other part of the third phase, plays a less active role. However, interactions take place between all actors in the network. Another difference between this part of the third phase and the other part is the fact that the main type of interaction in this part is characterised by confrontation and conflict. On the one hand the municipality and the province are of the opinion that the siting project has to be continued, whereas on the other hand neighbours hold the opinion that the realisation of this project has to be stopped. A compromise is not accepted by the parties.

We can characterise the relation between the VAM company and the neighbours of the proposed facility as a zero-sum game. The VAM company's benefits from siting the facility are directly related to the costs to the neighbours. The existence of a zero-sum game is an important factor in determining the interaction pattern. The distribution of resources among the actors is of great importance for the chances of success in this situation. An unequal distribution (the winner has many more resources available than the loser) is the only distribution that makes success possible. If the resources are distributed in an equal way among the actors, the success depends on the possibility to turn the process into a positive-sum game or to present it in such a manner. This is seen in the Wijster case. During one of the last stages in the decision making process the setting up of a compensation fund is proposed. The money from this fund should be available for waste prevention and recycling and for nature and environmental management. This means a flow of benefits to the neighbours if the incinerator is sited, which result in changing the zero-sum game into a positive-sum game. The proposal of the Agricultural Working Group (after losing any hope of stopping the project) to make a compensation arrangement if the incinerator is sited can also be seen in this way. Until now, no such arrangement has been made.

The interaction pattern between the VAM company and the organisations of neighbours can be characterised by confrontation or conflict. Both actors know that they cannot change the other' point of view and they both state that it is not very useful to deliberate with each other, because both actors stay with their own point of view. This situation develops later as the VAM company arranges public meetings to inform neighbours. An explanation can be found in the hypothesis that mere exposure to risk information results in negative reactions. Gutteling and Wiegman (1992:99) come to this conclusion in their research on the evaluation and experience of environmental risks.

Not much direct interaction exists between the municipality and the neighbours. One of the interactions is the public hearings. This interaction is characterised by confrontation or conflict. Both parties do not want to make any compromises. The municipality of Beilen takes a chance to turn the process into a positive-sum game by making the condition that the building of the incinerator has to be of benefit to local labour. We have to recall here the fact that the municipality also has an interest in increasing employment in its community. A similar situation holds for the interaction between the province and the neighbours. Direct interaction between the province and the neighbours is very rare. One of the occasions of direct interaction are the public hearings. This interaction is characterised by conflict and confrontation. Both parties have different opinions and neither of them accept a compromise. The province (as does the municipality) takes a chance to transform the process into a positive-sum game by making the condition that the building of the incinerator must benefit local employment. Here, too, we have to bear in mind the fact that the province also has an interest in increasing employment in its region. Schematically the network in this part of the third phase can be described as follows:

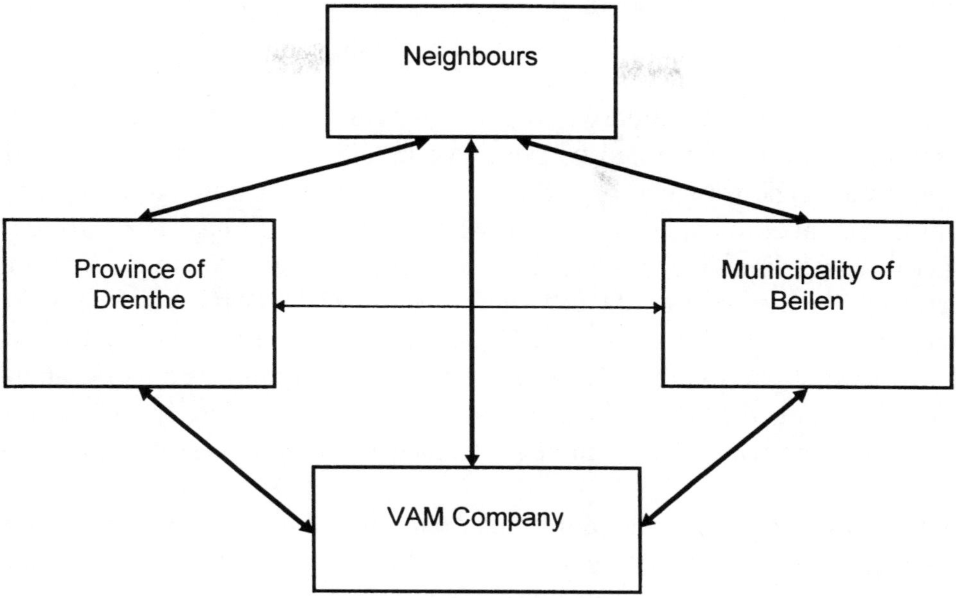

FIGURE 7.3B Interaction pattern in the objection part of the third phase of the decision making process

7.4 CONCLUSIONS

After analysing the dynamics of the Wijster case and its actors, we can draw conclusions on the factors that were responsible for the success of the decision making process. In order to do this we analyse the moments of choice that can be perceived in the decision making process. In this way factors of success can be derived from the choices that were made during the process. During the process several moments of choice appeared that were crucial to the development of the case. If we take a look at these moments of choice, we can ask ourselves questions like 'How do what choices lead to success?', 'Why has this actor at a certain moment reacted in this way, and what if he had not?' The answers to these questions can give us a clue to finding factors that have been essential to success. So, for all moments of choice we can analyse whether they were crucial to success.

A first step in this siting procedure was the initiative of the VAM company to build a waste incinerator. According to the VAM company they had no choice: building a waste incinerator was the only way to solve the problem of the ever-growing amount of delivered waste, by reducing the waste volume.

The VAM company proposed its plans to build a waste incinerator to the province. The province had two options: it could or could not agree with the project. The province found itself in a situation in which a ministerial guideline favoured incineration of waste over land filling. Furthermore, there was no direct incentive to the province to not agree to the proposal of creating an incineration facility within its own borders, which was in accordance with the ministerial guideline. As a result the province integrated this option into its plans by giving priority to waste incineration and indicating the VAM location as a possible site. The province first expressed this point of view in its waste plan and later in its region plan.

A following step was integrating this option into the municipal policy. To get this clear we can first find a clue in the socio-economic context. The region in which the incinerator was planned did not contain many sources of employment at the time the incinerator was proposed. As a result, siting a waste incinerator would have a relatively high effect on employment. Moreover, in case the incinerator was not sited, the continuation of a great deal of employment was in danger, because the continuation of the main employer in this region, the initiative taker on the project (the VAM company) would be endangered. This would imply the loss of one of the most important industries in this region.

But there were more essential factors. If the municipality did not agree with siting, and as a result the waste treatment at the VAM company in Wijster were to come to an end, the municipality risked becoming responsible for the present landfill. This could later on put an enormous pressure on the municipality's financial resources.

Another factor can be found in the location of the project. The city of Beilen has a long tradition of waste treatment. A landfill had been situated for decades at the proposed location of the VAM company. The waste is brought there by train from all over the country. This is why the infrastructure for delivering waste is already in place.

Another factor related to the location of the proposed facility can be found in the fact that the inhabitants of the municipality are used to waste treatment as an industry just like other industries. And a large share of the inhabitants benefit directly or indirectly from the waste treatment industry of the VAM company, either by having a job at the VAM company or by profiting from the facilities. On the other hand, this region always has had a low degree of pollution even at the time the incinerator was proposed. The VAM company also had a good record in this respect.

The factors mentioned above formed (among others) the reasons for the municipality of Beilen to agree with siting. As a result the municipality had the intention of integrating its point of view in its land use plans.

Until then the procedures were followed very strictly. The process was highly institutionalised. And actors interacted in a problem solving way.

At this point the opposers rushed into the process. There are many possible explanations for the opposers rushing in this late. One of them can be found in the fact that a few months before the opposition grew, the attention of the Dutch public was drawn to the 'Likkebaert' affair, named after a region. Dangerous levels of dioxins were found in the milk of cows that grazed in the neighbourhood of a waste incinerator in the Likkebaert polder. This resulted in closing down some Dutch waste incinerators until they met the prohibited levels of dioxin emissions. This could have made a profound impression on neighbours of a future waste incinerator in Wijster, who reacted with opposition to the siting of such a facility.

From this moment on the opposers took every opportunity to act against the siting of the proposed incinerator. But, after all, the procedure was already far along. There were not very many judicial opportunities left for opposers. Besides this, the opinion of the province and later the opinion of the municipality was fixed. The province had already agreed with siting by adopting the national priority on waste incineration in their provincial waste plan, and by indicating the VAM site as a location for waste incineration. This is why the opposers of the proposed site could not get support from these actors to stop the process.

Probably a main factor of a successful action against a proposed siting can be found in the possibility of opposers to get support from one or more public actors who play a formal role in the procedure. Because of this support they can gain a stronger position in the network of actors. One essential condition is that opposition can only be effective as the support is reached during specific phases during the decision making process. The opportunity is highest that public actors will give any support during an early phase in the process, because they still can take a position. During later phases in the process actors have probably fixed their opinion and are not going to change it. At the time the opposers in the Wijster case took their position it was too late to get any such support, because the public actors had already fixed their position. The decision making procedure was too far along. The opposers to the proposed siting started their opposition only at a moment when it would be very hard, if possible at all, to stop the siting process any longer, because some essential steps in the decision making process had already been taken. The provincial waste plan was already approved, and in it the priority had been laid on incineration of waste and the VAM site had been indicated as a location for siting a waste incinerator. This is why the opposers no longer had much opportunity to oppose under to the formal procedure. This can be seen as a factor in to the decision making process, whereas the opposers had had the opportunity to participate during citizens participation procedures concerning the provincial waste plan. The whole decision making process had a high level of institutionalisation. All formal steps in the process had been followed very carefully. So the later opposers had the opportunity to participate in earlier stages, but they did not use it at this time, basically because they were not yet opposers at that time. Because the procedure had nevertheless been carefully followed, the later opposers had no legal grounds to challenge the legality of these earlier steps.

Another factor that could have contributed to the success of the decision making process can be found in a factor that concerns the decision making process. From the start of the process, until the opposers rushed in, the pattern of interaction between the public actors involved and the initiative taker (the VAM company) had a problem solving character. The opposers, on the contrary, did not choose the problem solving way, but reacted in a conflictual and confrontantional manner. This alone could have formed a block to the new opposers. Moreover, they did not adjust to the culture that had existed up to that point: the opposers started accusing municipal as well as provincial deputies of having a personal interest in siting a waste incinerator because of their directorship of the VAM company. This not very tactful manner can be one of the reasons why the opposers were not drawn into the consultation by the other actors in order to solve the problem in a deliberative way. But they also did not use the earlier judicial opportunities to do so by themselves.

Were the decision making process not to have been as institutionalised as it appears to have been, maybe the opposers would still have had the opportunity to participate in a later phase of the process. If the process were not as highly institutionalised as this, one can in some cases fall back on earlier stages. Because in some cases these earlier stages give the opportunity to take them into account again because they do not have been handled as carefully as they might have been. So, the opportunity to disclose these stages is still there. However, the opposers probably could have gained most influence in participating this late, by acting in a problem solving way.

We can draw our conclusion by saying that the main factors of success resulted on the one hand from the proposed location for a waste incinerator, that contained not many sources of employment, not many pollution sources, and had a long waste treatment tradition. And on the other hand the character of the decision making process was due to the success: the late starting of opposition, the opposing in a conflictual way rather than a problem solving way, and a high level of institutionalisation, which provided no opportunity to disclose the former steps.

REFERENCE

Gutteling, J., and O. Wiegman, Beoordeling en beleving van milieurisico's: Overzicht van onderzoek naar ruimtelijke, sociodemografische, culturele en informationele determinanten, in: *Nederlands Tijdschrift voor de Psychologie*, 47, 1992. p 97-114.

8 A THEORETICAL FRAMEWORK FOR CASE STUDY ANALYSIS

BRUNO DENTE, PAOLO FARERI, JOSEE LIGTERINGEN

The six cases reported in the preceding chapters present stories on the decision making processes of six waste facilities siting procedures in different European countries. The studies include an analysis of several aspects of the decision making process, applying a common theoretical framework. This common theoretical framework, which is a rationale for case study analysis using a public policy analysis approach, will be clarified in this chapter. The benefits of a common theoretical framework can be found in the fact that it creates possibilities to compare different cases. Our conclusions, presented in Chapter 1, can be seen as a direct result of these benefits.

The theoretical research framework for case study analysis presented in this chapter can be seen as a framework in general: application is useful in reference to several fields of study. We can specify this general framework for the field we would like to study (as we have tried to do, for what concerns waste facilities siting: see Chapter 1.2). Every field has its own specific aspects in respect to the framework, making these aspects explicit enriches the general framework for that field specifically.

This chapter is divided into two sections. First, we describe the general theoretical framework for public policy analysis. The theoretical framework emphasises the need to discover the main elements of public policy analysis, such as are the problem, the actors and their resources, and the patterns of interaction between relevant actors, within the decision making process under study. These elements of public policy analysis are dealt with in Section 8.1.

Besides the usage of a general theoretical framework for case study analysis, the application of similar methods can contribute to the comparability of analyses of different decision making processes. In

B. Dente et al. (eds.), The Waste and the Backyard, 197–223.

Section 8.2 we concentrate on the actual execution of a case study analysis. In carrying out case study analysis we distinguish three phases: the construction of the chronology, the actors' analysis, and the patterns of interaction analysis. All three phases have their own methods and information resources.

8.1 ELEMENTS OF PUBLIC POLICY ANALYSIS

Decision making is a process of problem solving that reduces possible alternatives to a solution by choosing an option for realisation. *Policy making* also reduces the number of alternatives, in order to influence the actually stated policy. In this respect public decision making contributes to public policy making. A way to analyse public decision making is to use the elements of public policy analysis. Public policy analysis often starts with an analysis of the problem: what is the discrepancy between the desired situation and the actual situation for the actors involved? Above that: what is the overall problem? The same questions, analysing the problem, we refer to the analysis of a decision making process.

If we know what the problem is within the decision making process under study, we can transfer our attention to the actors: which actors have contributed to the process? Besides a listing of all the actors involved, we are interested in the actors' characteristics: what do they apply as their resources, and what are their particular motivations? The role that an actor plays in a decision making process is directly related to its motivation and its resources.

In order to complete the picture of the decision making process we pay attention to the interactions between the actors. Here we distinguish two aspects of interaction. One is the structure of the interaction. In this respect we refer to the shape of the policy network, resulting from mapping the interactions. Another aspect of interaction we distinguish is the character of the interaction, in the way that they interact in a specific manner, for example using negotiation and bargaining, or confrontation and conflict.

8.1.1 *The problem*

A problem can be characterised by the experience of a discrepancy between a standard or a desired situation on the one hand and the perception of the actual or expected situation on the other. Therefore, the experience of a situation as a problem situation depends on our standards as well as on our perception of the situation. Following from this definition of a problem, three aspects of the problem are important to mention here: the non-objective nature of a policy problem, its instability, and its consequences in terms of the analyst's role in the definition of the field of analysis. These aspects are of special importance in public policy analysis.

Because the problem is partially defined by a standard or desired situation, and the perception of an actual situation, the problem is not an objective datum, it does not exist in itself. Every actor defines the problem in reference to its own knowledge, resources, objectives, etc. Actors can differ in their standards as well as in their perceptions of the situation, and consequently in their problem definition. So the outcomes of the problem definition by different actors can be different and, in some cases, even conflicting. The central problem, then, needs to be 'defined', on the basis of the available information, resources and, last but not least, on the basis of the actors' objectives. Defining the central problem is an important part of a policy process. To reach an agreement among the participants about what the central problem is, is an important step towards the final success of the decision making process.

Furthermore, the problem is not stable. The situation that an actor desires, as well as an actor's standards and his perception of the actual or expected situation, can change over time. Consequently, the problem formulation can change in the course of the process, resulting from changing standards, or changing perceptions of the situation, which possibly can, among other factors, be caused by changes of the policy context, changes of the network of actors, or by changes of their objectives. The instability of the problem definition is an important aspect because it may contribute to a final success of the decision making process: the redefinition of the problem can be a way to find an agreed solution if it facilitates the integrability of different objectives.

Because the problem is not an objective datum, which moreover, is characterised by instability, in analysing decision making we need an *interpretation* by the public policy analyst. It is the analyst, on the basis of the different definitions of the problem that are posed during the process by different actors, who has to define what - from his/her point of view - is the central problem the policy is attacking. The importance of this interpretation becomes even more clear as we realise that the definition of this central problem is what we refer to in order to identify the 'boundaries' of the policy, in terms of the actions and interactions related to the treatment of that problem.

8.1.2 *Actors of the decision making process*

Besides defining the problem, in order to analyse the decision making process we need to focus our attention to the actors that contributed to the process. That is why in this section we first discuss the criteria for defining an actor in the decision making process. The actor's objectives seem to play a major role in defining an actor as such. Applying the actor definition we can list the actors involved in the decision making process. Besides a listing of the actors involved, in order to analyse the process we need to focus our attention on the characteristics of the actors involved: the actors' resources and their particular objectives. The role that an actor plays in the decision making process is directly related to the actor's objectives and its resources. After discussing the actor definition, and consequently the actor's objectives, we direct our attention in this section to the actor's resources and the roles actors can play during the decision making process.

Actors and their objectives. Every subject, individual or collective, public or private, participating in a decision making process and in that way applying a personal policy, is an actor in that process. By 'applying a personal policy' we mean here that the actor has an ambition to reach specific objectives within the process through specific means. By applying a personal policy the actor can contribute to solving, decreasing, or anticipating some (personal) problem.

An actor can be an individual as well as a collective subject. The condition for defining a collective subject participating in a decision making process as an actor is (only) the *internal homogeneity of its objectives* in any specific moment of time. If this condition is not fulfilled, we can determine several actors within this single collective subject. In this case we call the collective subject an 'arena' (see 8.1.3.), wherein different actors play their part. For example, a city council is an actor in a specific process only if it expresses unified objectives about the matter; otherwise it will be an arena in which different actors move. In this example political parties can be the different actors that make up the arena. Another example concerns conflicting positions within a political party. We will identify the groups holding different positions as different actors within the political party as the arena. The identification of the actors in the decision making process is a result of the analysis.

Within the actor concept, and consequently within the conceptual framework presented here, the definition of 'objective' is a central one. The definition of objective gives us the opportunity to articulate the framework concepts and to improve the interpretative tools of the model. However, we must note the difference between objectives and related concepts, like wishes, and ideals. The difference between an objective, and wishes and ideals in general is the fact that an objective goes one step further: the actor has decided to *try to realise* a wish or ideal.

The contribution to the reaching of its (global) objectives is a condition for an actor to participate in a network: the actor perceives that by participating it will get specific benefits. As with the definition of the problem, in defining an objective we have to notice that we are dealing with perception, not with an objective datum. It is the actor itself that, to a certain degree, creates the benefits. It is its ability to perceive the specific decision as an opportunity for reaching its global objectives, which moves the actor to decide to participate in the process. In this way we can trace an important resource for conflict resolution: by reframing the problem, or by representing the solution in a different way, we may create the possibility that some actors gain a better impression of the benefits the decision could afford them.

Several types of objectives can be distinguished. Here we introduce in particular a distinction identifying two types of objectives. Each actor can be characterised by the fact that it mainly holds one of the two types. The first type is *content* objectives. In defining content objectives we draw the reader's attention to the aspects of the solution under debate: the actor's content objectives result from the fact that it perceives that the debated option confers predominantly either costs or benefits. An actor with prevailing content objectives perceives that it will gain benefits if the implemented solution has certain characteristics, for example a park rather than a landfill, a nuclear power plant rather than a coal-fired power plant, or an automated light railway rather than an underground railway. The second type of objective we distinguish is *process* objectives: in this case attention is put on the actors' roles in the decision making process and not so much on the actual solution to the problem that will emerge. An actor holding mainly process objectives is mainly interested in consolidating the specific role it is playing during the process, for example because this will contribute to reinforcing its position within the relevant policy community. In general, process objectives are not explicitly expressed by the actors, but we can identify them by analysing actors' behaviour.

The distinction between 'content' and 'process' objectives is important in public policy analysis because it often happens that the actors change their position during the process, supporting different types of solutions at different moments of time, for instance. The reasons for these changes can be different. For example, the learning process induced by interaction, or the change of some aspects of the context in which the process is taking place, or the fact that the actor maintains process objectives. In the last case the change of position does not correspond to a change of objectives: it is the way to reach them. This actor is not interested in the specific solution to the problem, so if another solution can better fulfil its process objectives, the actor can easily shift to supporting this solution. Although objectives may change during the process, in general we can see that, if we define objectives in terms of content versus process, in most cases each actor remains stable in respect to its 'approach' to the process: throughout the course of events it will hold either mainly content or process objectives.

However, the distinction between process and content objectives gives us even more analytical opportunities. The type of objectives an actor predominantly holds (process or content) is related to some other characteristics of the actor itself. For example, we can say that actors holding mainly content objectives often have fewer resources in a bargaining situation, because they win only if a precise solution is implemented, whereas actors holding mainly process objectives may only be concerned for realising a solution in general. Furthermore, we can identify a direct relation between the type of objectives and the role played within the process. The mediator, for instance, is defined by the fact that he/she has no content objectives.

Moreover, the distinction between process and content objectives gives us the opportunity to see how the integration between actors holding different types of objectives in different phases of the process can facilitate the success of the decision making process. For instance, successful decision making is often the outcome of a process where an actor with content objectives raises the problem and proposes a solution, and an actor with process objectives drives the process toward the final agreement.

A second distinction in the objectives concept can be introduced, related to the one above; identifying *global* objectives and *specific* objectives. Let us coin a metaphor. We can represent a certain policy as a ring holding a sheaf of lines, coming from different points and going in different directions: each line corresponds to the chronological development of an actor (its 'life'). From the point of view of the actor, the policy we are analysing is a point along the line. We can define as 'global objectives' of this actor the direction that the actor wants to give to the line. These objectives have to do with general developments within the field wherein the actor moves, and (in general) not particularly with the specific decision making process in which the actor participates at a certain moment in time. Global objectives do not depend on (nor are the result of) the specific policy we are analysing: on the contrary, these objectives are a condition for the participation in the process. An actor decides to join a policy network to a specific process if it perceives that this could be an opportunity to develop the line in general in the desired direction. Conversely, the 'specific objectives' are the application of the global ones to the process we are analysing: by holding a certain position within the network, the actor perceives that it is proceeding along the line in the direction that it wants to. In this way, aiming at specific objectives contributes to the realisation of the global objectives. Sometimes joining a specific process results in a change of the global objectives. This can happen when, as an outcome of the interaction in a specific process, the relation between the ring and the lines goes in the opposite direction. Consequently, participation in a decision making process can result in a reorientation of the line, in a direction which is different from what the actor decided before joining the network, which means a change in the global objectives.

The 'distance' between global and specific objectives can vary, depending on the actor's characteristics. That is why we have to make some necessary differentiations with respect to distinguishing global and specific objectives, in particular where we said that global objectives have nothing to do with a specific decision making process in which the actor is joining. Sometimes the distance between global and specific objectives can be so close that the difference between global and specific objectives may disappear. For instance, in the case of a local group, constituted to oppose the realisation of a waste facility, the distance is very small because the actual existence of the actor depends on the existence of that specific decision making process. A completely different case is when actors are involved in a broader range of activities. In that case the distance between global and specific objectives can be substantial. Examples of actors involved in a broader range of activities are the Department of the Environment, and a private company specialising in waste treatment at a national or international level.

Finally, we can specify the relation between the two types op objectives we have distinguished: content and process objectives versus global and specific objectives. Because both content objectives and process objectives refer to the specific decision making process in which the actor participates at a given time, these two can be identified as specific objectives. The result is that we can distinguish on the one hand the specific objectives, which can subsequently be divided into content and process objectives, and the global objectives on the other. In short, when we analyse an actor's behaviour in a specific decision making process, we identify its specific objectives (including content and process objectives); but to understand why it holds its position, we have to take its global objectives into consideration. The relation between the two is a key element in public policy analysis.

Actors and their resources. Every type of behaviour requires resources. In participating in a decision making process every actor uses different types of resources to reach its objectives. There is not a hierarchy of types: one actor, for instance, can balance the lack of one type of resources (such as economic resources) by using another type of resources (such as political or strategic resources), resulting in a win in the confrontation with another 'richer' actor. The 'power' (weight) of an actor is therefore determined by the total amount of different resources, and by its ability (which is a resource in itself) to use them within the process.

Four main types of resources can be identified. The first type is *economic* resources, in terms of the amount of money that can be mobilised during the process. Money can be mobilised in order to be exchanged with other actors and/or to buy other types of resources. The second type is *political* resources, i.e. the consensus an actor is able to gain to legitimise its position within the network to the other participants. Political resources can be defined in terms of following, or sympathy, or support from the media, but also in terms of confidence. Confidence is one of the bases for almost every interaction, and can therefore be seen as an important political resource. The third type is *legal* resources, in terms of the advantages the existing laws can give to an actor, such as the right to be consulted, or the requirement to reach an agreed solution. It is important to note that this is one resource among the others. The legal status does not, in itself, define the power of an actor. It is rather its use as a resource that can put the actor in a better position within the network. The fourth type is *cognitive* resources. Different types of cognitive resources can be identified. *Scientific knowledge* is data and/or models produced by experts (consultants or bureaucracies), and can be used by an actor to influence other actors' behaviour. We must pay attention here to the identification of the actors, because in some cases the expert itself becomes an actor of the process (usually holding process as well as content goals). *Interactive knowledge* is knowledge produced by the actors during the process regarding the process itself, such as information about the structure of the policy network, and other actors' behaviour, and resources. Interactive knowledge is very important because it is the informative base for another cognitive resource, viz., *strategy*. We define 'strategy' as the ability of an actor to understand other actors' behaviour and, by anticipating it, to turn it to its advantage. Within a strategic approach, some elements of a decision making process - otherwise perceived as constraints - can be used as resources. This is, for instance, the case with time. Both the lack and the availability of time are not resources in themselves, but can be used as resources within a strategic approach.

We should note here that some resources may be exchangeable and others may not. If we look, for example, at money as an example of an economic resource and at authority as an example of a legal resource, we see that money is in general exchangeable and most of the time authority is not. The exchangeability of resources has its consequences for the power an actor derives from its resources. If resources are fully exchangeable, and resources are evenly distributed among the actors, the network can be characterised by a mutual dependence of the actors. On the other hand, if resources are not exchangeable, and again resources are evenly distributed among the actors, this does not consequently imply the existence of mutual dependence. A lack of exchangeability can give an absolute power basis, and consequently a certain independence to an actor.

Finally, we should make one further remark here. The power of an actor is not only determined by the total *amount* of different resources, and by its *ability* to use them within the process, it is also determined by whether an actor *actually uses* the available resources if it has the ability to use them. This is what we see, for example, in the case of authority. Authority can be a power basis only if an actor uses this as a resource during the process.

Actors and their roles. Since we know that factors like the problem the actors perceive and the actors' resources play a major part in the analysis of the decision making process, we can also direct our attention to the roles actors play in the process. The role that an actor plays is directly related to its objectives and its resources, and consequently to its position. However, this role can be totally independent of its *formal* institutional position. Of course some relations can be identified. For example, the formal institutional position the actor holds can determine some crucial available resources, but at the same moment, the available resources can determine the role an actor plays during the decision making process.

Here we describe some of the possible roles the actors play within the network, apart from their institutional status. The list is not complete. One of the goals of the analysis of a decision making process is to assign a role, as actually played, at least to the main actors involved.

The Promoter. The promoter is the actor that plays the central role during policy formulation: it raises the problem, activates the subjects responsible for its treatment, and in most cases proposes a solution. It has mainly content objectives: it is interested in the fact that the problem is perceived and attacked. In many cases the promoter uses cognitive resources, both in terms of scientific knowledge (it knows the problem and the solutions) and in terms of strategic resources (it is able to activate the network).

The Director. The director is the actor most interested in reaching an outcome, overcoming the possible problems of process development, and the unintended consequences of interaction. When the decision problems mostly depend on the lack of interest of the actors responsible for implementation, it will be more likely to find as director an actor holding mainly content objectives (an extension of the promoter). In a process characterised by a high level of conflict, by contrast, it will be more likely to find an actor with mainly process objectives. In both cases, strategy is a very important resource for the success of the director.

The Opposer(s). No explanation is needed here: the opposer is an actor acting against the implementation of a defined solution. This role can be played both by actors holding mainly process objectives or content objectives. We can have as opposer the minority coalition in a city council, strategically using the contingency to provoke a political crisis. Or we may find as opposer a local group of residents, fighting against the implementation of a certain project that confers costs on them.

The Mediator. The mediator is a subspecies of the director. This is typically an actor with mainly process objectives. It tries to drive the process toward an agreed solution to legitimise its presence as an actor in the process, for example in order to become a stable actor in the policy community regarding that policy area. Not holding well-defined preferences about the alternative solutions is a decisive condition for mediating between actors holding conflicting interests.

The Gatekeeper. The gatekeeper is an actor which, often on the basis of legal resources, or of cognitive resources, is able to block the advancement of the decision making process. Without its consensus the process cannot proceed toward its termination. In most cases the gatekeeper holds mainly process objectives: its participation is the result of its stable position in the policy area of which the decision making process is part. So the gatekeeper's objective is to consolidate its privileged position and to prevent the adoption of solutions involving costs at its charge.

The Filter. The filter is often a case of an induced actor. It does not hold specific interests about the issue, or at least not enough to justify its participation in the network as an autonomous decision, but becomes the carrier of another actor's requests (under pressure) to a third participant. The typical case is that of a city council representing the position of a local group to a central government body.

8.1.3 *Interactions within the network*

In completing the overall picture of the decision making process the interaction between the actors plays a major part. First we have to notice that there is some issue to be decided on in public policy that makes actors interact with each other, and in that way constituting the network in this specific process. The issue that is to be decided upon we refer to as 'the stake' of the decision making process.

Within the interaction concept we can distinguish two network characteristics that can help us to explain the decision making process: the prevailing culture it holds and the structure of the network. The cultural aspect of the interaction we call the 'interaction pattern' in the policy network. In this respect we could think of the way in which actors interact, for example in bargaining or more confrontation-like ways. A related concept is the policy arena, which refers to the setting where the interactions take place. The structure of a policy network can be derived from the existence of interactions between the actors within the network, in this respect answering questions like: do all actors interact with each other, or in other words, or do actors only interact with one central actor in the network?

We should notice here the difference between the policy network in general and the network with respect to the specific decision making process we study. A policy network with respect to some policy field can in general be seen as durable, but its characteristics may differ slightly with every decision making process the network is involved in. In this respect we distinguish between issue networks and policy communities. The difference between these is especially apparent with respect to network characteristics like the pattern of interaction and the structure of the network. Here, if we mention characteristics of the network we refer to the characteristics of the network in the specific decision making process we study.

The stake. The stake of a decision making process can be defined as *what is going to be decided about in a public policy*. The stake is, consequently, what the actors refer to in order to evaluate the possible costs and benefits the policy would distribute to them. In this way, the stake is the binding factor that brings the actors within this specific decision making process together in terms of interaction. From the point of view of each actor, the perception of what the stake is can be different, depending on the part the actor is playing within the network and in the policy area.

Let us take as an example the case of a decision making process which is part of a policy area characterised by the presence of a well structured policy community. In this process it will be very likely to find 'occasional' actors - their presence being determined by the specific problem treated in that process (a local group of opponents, the city council, the landowner, etc.) - and 'stable' actors - the ones participating to the policy community (the ministry, a public agency, a private company specialised in the matter, etc.). While the first interact with all the others only during the process being analysed, the latter interact in different processes (before, after, and maybe at the same time). These two types of actors' perception of what the stake is can be different: more limited to the specific decision that is going to be taken for the occasional ones, more concerned about the consequences that the decision can have on the structure of the policy community for the stable ones. Consequently, depending on the perception of the stake, the actor defines its (content or process) objectives.

The stake determines the basic aspects of the patterns of interaction of a decision making process. We can very briefly introduce here some concepts taken from 'game-theory'. A policy can be represented as a *zero-sum game* when the stake determines as an outcome a distribution of costs and benefits so that it is easy to find winners and losers in the process. In this type of game the condition for giving benefits to an actor is to distribute costs to another, in this way creating a *win-lose situation*. By contrast, we will have a *positive-sum game* when the outcome of a process distributes benefits at least to all the main actors involved, creating *win-win situations*. The nature of zero-sum versus positive-sum games in a process directly influences the patterns of interaction and, in relation to the distribution of resources among the participants, determines the process's chances of success. The success of a zero-sum game is possible only if the winner has a lot more resources than the loser. If the resources are well distributed, on the contrary, success depends on the possibility to transform or at least represent the process as a positive-sum game. However, we have to notice here again the exchangeability of the resources. As we have seen, the extent of exchangeability of resources can influence the degree to which the actor actually can derive power from its resources.

The stake is not necessarily stable in the course of the process, and it can be modified by the actors themselves in order to reach their objectives. The stake can change just as much as the objectives of an actor or his problem definition can change during the process. The stake is not only influenced by internal changes within the actor (like its problem definition, or its objectives), but also by external changes, which might even be manipulated by other actors in order to do so. In particular, intervening in the stake (by increasing or reducing it) is often a way to change a process from a zero-sum to a positive-sum one. Compensation is an example of this strategy: distributing extra benefits to the losers by increasing the stake, obtaining a positive-sum game. Again, we have to notice the importance of exchangeability. Compensation may only be possible if there are exchangeable resources that can be distributed to the losers.

Patterns of interaction. The culture of a network of actors within a specific process can be identified by the pattern of interaction. A pattern of interaction involves the way in which actors approach each other and subsequently continue their interaction. Patterns of interaction can also be seen as strategies. Applying these strategies, the actors try to reach their objectives by influencing the decision making process. Three main patterns of interaction can be identified. The first one is *confrontation*, or conflict. In this case the main actors interact, each one trying to prevail on the others, refusing the possibility of compromises, and consequently maintaining the conflict situation. Confrontation or conflict only follows from situations in which both actors (or at least the actor that maintains the conflict situation) have the impression that sooner or later their own resources will be sufficient to win over the others. The second type of interaction pattern is *bargaining*. We still have conflicting objectives here, but the actors perceive that none of them has enough resources to win over the others. Bargaining usually means that both actors partly give in on their wishes. The third pattern of interaction is *problem solving*. A problem solving approach is not only adopted when the actors have the same objectives. Actors may have objectives that do not at all interfere with each other's fulfilment (a *win-win situation* actually exists) or actors may perceive that their objectives go in the same direction and are integrable in that way. In this case the main actors look for a collective definition of the solution.

The different patterns of interaction depend on the actors' perception of the distribution of resources. A confrontation pattern is often the result of the concentration of crucial resources on few actors, perceiving that they could win without any need for a compromise. In these cases the crucial resources are often unexchangeable, or the remaining actors perceive them as not likely to be exchanged by the actors. This is why bargaining is perceived as out of the question. The other two patterns (bargaining and problem solving) are the result of a broader distribution of resources. In the problem solving pattern resources are so diffused that the only way to get the solution is to co-operate, more than to obtain consensus from some of the participants. This presupposes that resources are exchangeable, because otherwise there is no need to bargain.

The pattern of interaction within a policy network can change during a decision making process. Actually, once again, what determines the strategy is the actors' perception of the process. The perception of the available resources and of the exchangeability of the resources is a crucial one within the choice for a specific interaction pattern. Because we have to do with perceptions, strategies may work out wrong for the actor involved as a result of an unrealistic perception. A change in the interaction pattern can be the result of an actor's misperception. For example, an actor can presume to have enough resources to impose its solution to the other participants (this perception leading to a confrontation pattern), and then realise that the opposers are stronger than he imagined. The outcome of this situation can be blocking of the process, or a change towards a bargaining pattern.

Besides policy networks we distinguish policy arenas. Where policy networks concentrate on the entirety of actors and their relationships, the *policy arena* focuses on the setting in which these interactions take place. A policy arena is where (some) interactions take place. A typical example of a policy arena is a city council where different party groups interact. More than one arena can be found in a policy network. As we did with respect to policy networks, we can characterise arenas in terms of their predominant culture. This results in four main types of arenas:

- *technical arenas*, where the interaction is mainly among professionals, and the discourse is centred around the technical aspects of the decision;
- *political arenas*, where the participants are mainly the political parties;
- *social arenas*, where the interaction involves social groups, sustainers and opposers; as in political arenas, an important role can be played by the media, and the public opinion;
- *institutional arenas*, where the interaction is among the bureaucracies of different bodies of the public administration.

The structure of policy networks. Besides the pattern of interaction, we can identify the structure of a network as an aspect of the interactions within the network. Like the pattern of interaction, the structure of the policy network can contribute as a factor to the explanation of decision making processes. A policy network is *the whole cast of actors participating in a decision making process, and the relations among them.* As we saw, some durability is supposed within the general network with respect to a certain policy field, but over time a policy network is not necessarily stable in its composition over time; furthermore the composition of the general network may differ slightly with every specific decision making process: as often happens, new actors can enter the decision making process during its development, and some actors can leave the network for several reasons, for example if they lack relevant objectives. Besides the composition of the network, the relations structure among the actors within the network can change. Changes in the network structure can result from changes in the interactions, or from changes in the composition of the network itself.

We can represent the structure of a policy network by mapping the actors involved and the relations among them, identifying the existing connections, their intensity, and the changes in the course of the process. Different types of policy network structures can be identified. A non-exhaustive list includes, for example, a *star structure*. In this case a 'central' actor can be identified, which is the only one holding relations with all the other participants, that are not interacting with each other. A policy network can furthermore be characterised by an *axis structure*.

Here we have a limited number of main actors, connected one another by the most deep relations of the network. Each of the main actors relates to some of the marginal ones. The 'axis' can be, as a variant, a 'triangle', a 'square', etc. What is important here is that the relations among the actors have different levels of intensity. Another example of a policy network structure is the *total interaction structure*. In this case it is difficult to identify a hierarchy of actors and of relations. Every actor interacts with all the others.

It is important to note that categorisation does not always clarify the decision making process, but sometimes it contributes to a better interpretation of the process. Mapping the policy network, representing the interaction with a scheme like the ones described above, can be very useful, for example, in identifying actors' roles and the success factors of the decision making process.

8.1.4 *Conclusions*

In analysing public policy we consider some elements to play a major role. First, a definition of the problem is a crucial step. Actors can define the problem in different ways at different moments during the decision making process, and the analyst has to interpret what the central problem is in the decision making process studied. The importance of defining the central problem can be found in the fact that the central problem determines the boundaries of the analysis: aspects that are related to this central problem or its solution are the subject of the analysis.

A second element of public policy that needs our attention is the actors participating in the decision making process. Actors can be characterised by their own objectives, resources, and roles. The actor's objectives and its resources determine for the most part the role the actor plays within the decision making process. The actor's characteristics can change during the decision making process, and in that way can influence the outcome of the process.

A third major element of public policy analysis concerns the interactions between the actors joining the decision making process. In this respect we need to notice the fact that the actors can also join other processes, simultaneously as well as prior to the process being studied. Although these external interactions are not the subject of study, they may influence the developments in decision making. In analysing the interactions we can pay attention to the character as well as the structure of the interactions. The character as well as the structure of the interactions may determine the success of the decision making process.

After determining the elements of public policy analysis in general we can complete them for specific fields of study. In the next section we complete the elements of public policy analysis for the field of siting waste facilities.

8.2 METHODS AND TECHNIQUES FOR CASE STUDY ANALYSIS

Applying general guidelines for case study analyses, in research projects that contain several cases, gives the opportunity to guarantee the homogeneity of the analysis, and consequently the comparability of the results of the different case studies. These general guidelines for case study analysis are presented in this section. The reconstruction of decision making processes is based on three main phases, to be developed in chronological sequence. Each phase is supposed to be autonomous, using different sources and different techniques. At the same time, the completion of the previous phase is a condition for the development of the subsequent one. Notice here that distinguishing phases is only a suggestion in implementing research activities, which does not automatically mean that we prefer to present the results of case study analysis in three divided sections, following the three phases of case study analysis. There can be different reasons for following another order, or mixing up the presentation of the results. The results of the case study analyses we presented in the previous chapters give an example of mixing up results that are drawn from different phases in the case study analysis. The three phases of reconstructing decision making processes we distinguish are:

- the construction of the chronology of the decision making process;
- the analysis of the actors involved;
- the analysis of the patterns of interaction and the definition of success factors.

In this section a brief description of the objectives, techniques and sources is given for each phase.

8.2.1 *Constructing the chronology*

The construction of the chronology as a first phase in reconstructing the decision making process is dedicated to the analysis of the 'story' defining the decision making process, i.e. the identification of all the events which, in the judgement of the analyst, in some way conditioned the development of the process, from the initial policy formulation to its termination.

Objectives. The objective of the construction of the chronology of the decision making process is to define the policy framework (the boundaries of the policy), under two different dimensions. The first dimension is *historical.* Since the decision making process does not necessarily coincide with the formal procedure, the objective of this phase is to identify the course of events we consider part of the process, in order to explain its outcome. These include, apart from the formal acts, a very broad range of actions, such as the change of local government coalitions, the constitution of local opposition groups, the organisation of a scientific conference, the publication of a research report, the raising of a conflict about a proposed solution, and the effects of other decision making processes modifying the objectives of some actors.

The second dimension is related to the *policy network.* Identifying the events corresponds to identifying the actors involved in the process. So this phase, apart from the chronology, produces what at this point of the analysis is simply a list of the actors involved. This list is a condition for the development of the second phase.

Techniques. Usually the chronology is constructed starting from the *outcome:* the event more easily known at the start of the research, and proceeding backwards, identifying the connected events until one reaches the moment when the problem is raised. It is very important to adopt, at least in the beginning, a broad criterion for the selection of the events, reducing the risk of losing important aspects of the process, remembering that it will always be possible, during the subsequent phases, to exclude information that is not relevant.

Here we consider the chronology as an instrument used by the analyst for the development of the other phases. For this reason the chronology does not contain any interpretation of the process: just the raw facts, as they are described by the source. It is important to note that there is no intention towards 'objectivity' in this. The interpretative action by the analyst in the construction of the chronology is the selection of the events. The chronology can, up to the completion of the three phases, be treated as a rough document: just the chronological sequence of the events (separated from one another, identified by date), containing all the information considered useful for the analysis. Furthermore, the chronology is an open document, to be integrated, if necessary, with additional information in the course of the analysis, for it often happens that other important events are identified after the completion of the first phase. It could be useful for the analyst to indicate the source of information for every event, because it could be necessary - in the following phases of the work - to go back directly to the source to integrate information.

In order to favour an external reader, only after the completion of the whole case study analysis is it possible to change the style of this phase, describing the story in a more narrative way, and excluding all the information that is not relevant to understanding the development of the process.

Sources. Three main types of sources are used for constructing the chronology. The first source is *the press review*, which is the collection of all the articles published by local, national, political newspapers and magazines and by specialised reviews, containing information about the decision making process. Usually this is a very important source, especially if the process is characterised by a high level of conflict, because it gives us a more objective picture of events. It is important to collect all the available articles, because this contributes materially to the analysis. For example, it can be useful to compare how the same event is described by different (local, national, political) newspapers, or how different types of articles (editorials, reports, etc.) give different types of information about the same event.

The second source for constructing the chronology is *the (formal and informal) documents* issued by the actors involved and in some way relevant to the process. We could, for instance, think of acts, minutes of meetings, research reports, or pamphlets by actors like the central government, city councils, political parties, neighbourhood associations, environmentalist groups, or consultants. In order to construct the chronology we are interested both in the documents strictly related to the process as well as in those that can be useful to understand how every actor defines its more general objectives. In this phase the documents are used to identify the events, and to collect information about them. Often these documents contain a summary of what has happened before, helping in the construction of the chronology.

The third source in constructing the chronology is *in-depth interviews* of observers or informants of the decision making process. These interviews have the objective of integrating the different sources and of finding the missing parts of the process. Consequently, the interview should be based on the reconstruction of the story, as the observer knows it. It is better to have a basic knowledge of the process before interviewing the observers, to facilitate the interaction and to maximise the results. On the other hand, the interview must be very open, letting the observers say whatever they know.

The observers or informants can be selected according to several criteria. Obviously, their knowledge of the process is an important criterion. Beyond that, observers and informants should be real 'observers' and not actors: they should not have played any important role within the decision making process. Typical examples of observers are: the journalist(s) responsible for the local news; local militants of political parties (with a marginal role within their organisation); bureaucrats without specific responsibilities related to the process; members of the professional community with knowledge of the process because the content of the decision is related to their work. Above this we have to notice that the more frustrated (or the more satisfied) the observer is, the more he will talk, but in that case we will have to pay attention to selecting information from the judgements.

Finding a good observer or informant is a very important step in the economy of the work. Usually this observer has already collected articles and documents about the process, and in this way we have the easiest access to sources that are otherwise quite difficult to find. Almost every process has its key observer (the one that knows the story very well right from the start, and with shelves full of documents and articles), the problem is to find him.

8.2.2 *Actor analysis*

The most important result from the construction of the chronology is a listing of the actors involved. Piece by piece we can make up from the chronology a picture of every actor involved. Now we can construct a monograph for each actor (or at least the most important ones) participating in the decision making process. A monograph may contain all the useful elements for the overall analysis.

Objectives. The objective of the actor analysis is to describe the behaviour of the main actors involved in the decision making process, and the factors determining it. For a more detailed description of these factors we refer to section 8.1. In the second phase we should be able to specify the following aspects for each actor. First, we need to define *the problem* from the point of view of the actor, and the possible changes (redefinitions) in the problem perception of the actor during the course of the process. Related to the problem perception is the stake of the decision making process. That is why, second, we need to know what, from the actor's point of view, *the stake* of the process is. In order to get an impression of every separate actor we, furthermore, need to focus our attention on the *objectives* the actor holds within the decision making process. Again, we must notice that the objectives may change during the course of the process. Within the objectives concept we can specify the general and specific objectives on the one hand and the content and process objectives on the other. A fourth characteristic of the actor we need to specify is the availability of *resources* to the actor. Within the resources concept we distinguished economic, political, legal, and cognitive resources. Our question here is which resources the actor mobilised and which ones he used during the course of the decision making process. Finally to get an overall impression of a specific actor we can try to draw a picture of the actor's perception of *other actors' behaviour*. The actor's perception of other actors' behaviour possibly contributes to a modification of the actor's own behaviour, its objectives, and strategies, for example.

Techniques. The actor analysis phase is based on the development of *informal, in-depth interviews* with the main actors identified in the chronology (usually one interview for each actor). This type of interview has no set questionnaire: the questions are open ones, designed to encourage the respondent to talk freely around each topic. The structure of the interview is based on two different levels of questions. The first level is a set of open questions: this is the stable part of the interview, applied in the same way to all the actors. Examples of these questions are: *What was the problem the process was attacking; Describe the 'story' of the decision making process; Which were the constraints that made it difficult to reach a positive outcome; What is their perception of the elements that facilitated reaching a positive outcome.*

The second level is a set of specific questions, related to the actor interviewed, which should help the interviewer to conduct the discussion about each of the topics of the first level (specifying some aspects considered relevant in each case), and in reaching the objectives mentioned above.

The interview should treat all the elements of the monograph, even if some of them cannot often be posed in an explicit way, such as objectives and strategies. These aspects should be deduced by the discussion around the most general themes that can be found in the description of the story, 'strategically' using the second level questions.

Part of the interview should be dedicated to the construction of an 'identity card' for the actor. The discussion here concentrates on the actor itself, not on the process, identifying its structure, its range of activities. This part of the discussion has the scope to gain general information about the actor and should be conducted in such a way as to identify its global objectives and the available resources mobilised in the course of the process.

It is important to note that the objective of the interview is to collect information about that actor, and not the others, nor the story. What we get when the respondent talks about the other participants is its representation of the actors, and consequently this should not be a source for other monographs. Obviously the interpretation is based on the comparison of the interviews, but what we are looking for - in this phase - is *not an objective and unique 'truth'*. If different actors give different descriptions of the process, it does not mean that someone is 'lying' and consequently that the results of the interview are useless. On the contrary, what we are looking for are precisely the different representations of the process, in order to identify how each actor perceives the process and its role within the network.

It is always important to end the interview by asking the actor to indicate a *list of subjects* which, in its judgement, could be useful to interview. This list should correspond to the actor's perception of the composition of the policy network, and it is useful both for the interpretation of the actor's behaviour and to verify whether the analyst's list of actors is complete.

Sources. Apart from the main source, the interviews, the ones collected in the first phase should be used here to integrate information. For example the positions by the actor as described by the media, and the documents. The interview itself can be a good opportunity to integrate the collection of documents. Comparing documents (issued during the process) and the positions expressed in the interview can be useful to overcome the possibility of a 'cognitive dissonance' effect, i.e. the unintentional modification of some aspects of the actor's behaviour in the process as described by the actor itself, as a consequence that of speaking with hindsight, knowing what the outcome has been.

8.2.3 *Interaction analysis and the definition of success factors*

The last phase consists of the construction of the overall interpretation of the case study, on the base of the chronology and the actors' analysis.

Expected output. The overall interpretation of the decision making process includes at least five elements. First we need a definition of the 'real' problem attacked in the process by the point of view of the analyst. The identification of the roles played by the different actors in the course of the process is a second aspect that needs our attention in the overall analysis. Furthermore, a description of the patterns of interaction can give an idea about the composition and the structure of the network, the existing policy arenas and the modes of interaction (confrontation, bargaining, 'problem solving'). The transformation in the course of the process should be indicated for each of these elements. A fourth element of the overall interpretation consists of the identification of the process phases.

Phases of the decision making process can be distinguished referring to four variables. First every phase has its own definition of the problem. We must notice that this is the problem definition according to the point of view of the analyst. Second, we can distinguish phases within the process on the basis of the actors participating in the process. The composition of the network often changes per phase. Third, the structure of the network can play a role in distinguishing separate phases. In respect to this we must pay attention to the fact that the structure of the network can change, even if the actors remain the same, because the relations among the same participants can change. Finally, in distinguishing phases we need to focus on the prevailing mode of interaction (confrontation, bargaining, problem solving). Representing these variables, running through the phases of the decision making process, can be done by filling in the following chart.

	Phase 1 **Since: ....** **To: ....**	**Phase 2** **Since: ...** **To: .....**	**Phase n** **Since: .....** **To: .....**
Problem definition			
List of actors			
Network structure			
Prevailing modes of interaction			

The chart can be filled in by identifying a new phase every time a change even in just one of the four variables is discovered. A graphic representation can be useful to represent the structure of the network.

Finally, the identification of the success factors can be seen as a fifth element of the overall interpretation of the decision making process. Taking into consideration the results of case study reconstruction as a whole, the analyst should express here his/her judgements about the elements of the process itself that contribute in a more important way to explaining why a positive outcome has been reached.

Sources. We do not have to collect new sources in the third phase of the case study analysis, because this phase (patterns of interaction analysis and the definition of success factors) is the result of the analysis of the research reports outcome of the previous phases. If necessary one can go back to the related sources of these previous phases.

8.2.4 *Conclusions*

In order to create the possibility to compare different cases we need to be sure that these cases have been studied by applying the same methods and techniques. Here we present some guidelines for executing case study analysis. We distinguish three phases in case study analysis, which have their own methods and techniques. Within the first phase of case study analysis, the construction of the chronology, we define the political framework in historical respect and regarding the policy network, mainly using written sources and the information an observer provides. The second phase in analysing a decision making process consists of the actors' analysis. Within the framework of the actors' analysis we determine several aspects according to the actors involved: the problem perception, the stake, the objectives, the resources, and the perception of other actors' behaviour. In this phase we mainly use as a source informal, in-depth interviews with representatives of the actors involved. Finally, we distinguish a phase that concerns the analysis of the interactions and the determination of factors that may explain the outcome of the decision making process (success or failure). In order to execute this final phase in analysing the process we have to focus on the central problem, the actors participating in the process, the network structure, and the prevailing mode of interaction. These variables can help us distinguish different phases of the decision making process. Explaining the outcome of the decision making process results from an interpretation of the overall picture of the process that has developed by now. Because this final phase is characterised by interpretation, we do not need new sources.

8.3 CONCLUSIONS

Comparing different case studies is a difficult job, even more if the different studies are executed by different researchers. Every researcher has his/her own way of studying a specific case. In order to reach more comparability among case studies we have developed a theoretical framework that can be applied in general to different case studies done by different researchers. This general theoretical framework focuses on several elements of public policy analysis: the problem, the actors involved, and the interactions within the network of actors involved. Within the general framework each element associates with several aspects that have to be noticed in carrying out a case study analysis.

If we apply the general framework to a specific field of study we can in some respect complete the general elements, and associated aspects, for this specific field. In this book we paid attention to the field of siting waste facilities. This is why in this chapter we complete the general framework concerning decision making on waste facilities in particular. In doing this we distinguish several variables that may specifically in these cases contribute to explaining the developments within the process.

Besides applying a common theoretical framework, we can contribute to the comparability of different case studies by applying the same methods and techniques in executing the research. The guidelines concerning the methods and techniques of case study analysis presented in this chapter can help us in this respect. They guide the execution of the research through different research phases, recognising different research objectives, techniques, and sources.

Applying the general theoretical framework as well as the guidelines for case study analysis with respect to different cases may result in a high level of comparability, and consequently in an additional value of the available material.

Environment & Management

1. O. Kuik and H. Verbruggen (eds.): *In Search of Indicators of Sustainable Development.* 1991 ISBN 0-7923-1249-X
2. R. Janssen: *Multiobjective Decision Support for Environmental Management.* 1992 ISBN 0-7923-1908-7
3. R. Janssen and M. van Herwijnen: *DEFINITE. A System to Support DEcisions on a FINITE Set of Alternatives.* Software with User Manual. 1994 ISBN 0-7923-2696-2
4. D. James: *The Application of Economic Techniques in Environmental Impact Assessment.* 1994 ISBN 0-7923-2721-7
5. P. Glasbergen (ed.): *Managing Environmental Disputes: Network Management as an Alternative.* 1995 ISBN 0-7923-3034-X
6. L. Reijnders (ed.): *Environmentally Improved Production Processes and Products: An Introduction.* 1996 ISBN 0-7923-3786-7
7. E. Beinat: *Value Functions for Environmental Management.* 1997 ISBN 0-7923-4662-9
8. B. Dente, P. Fareri and J. Ligteringen (eds.): *The Waste and the Backyard.* The Creation of Waste Facilities: Success Stories in Six European Countries. 1998

ISBN 978-90-481-5021-2

KLUWER ACADEMIC PUBLISHERS – DORDRECHT / BOSTON / LONDON

SPRINGER NATURE

GPSR Compliance

The European Union's (EU) General Product Safety Regulation (GPSR) is a set of rules that requires consumer products to be safe and our obligations to ensure this.

If you have any concerns about our products, you can contact us on ProductSafety@springernature.com

In case Publisher is established outside the EU, the EU authorized representative is:

Springer Nature Customer Service Center GmbH
Europaplatz 3
69115 Heidelberg, Germany

Kolibri 360 GmbH
Ferdinand-Jühlke-Straße 7
99095 Erfurt, Deutschland
produktsicherheit@kolibri360.de